D1124589

Reflections on the Musical Mind

Reflections on the Musical Mind

AN EVOLUTIONARY PERSPECTIVE

Jay Schulkin

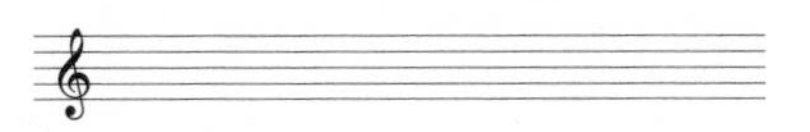

PRINCETON UNIVERSITY PRESS
PRINCETON AND OXFORD

Published by Princeton University Press, 41 William Street,
Princeton, New Jersey 08540
In the United Kingdom: Princeton University Press, 6 Oxford Street,
Woodstock, Oxfordshire OX20 1TW

press.princeton.edu

Library of Congress Cataloging-in-Publication Data

Schulkin, Jay.
Reflections on the musical mind : an evolutionary perspective / Jay Schulkin.
pages cm
Includes bibliographical references and index.
ISBN 978-0-691-15744-3 (hardcover)
1. Music—Psychological aspects. 2. Music—Origin. 3. Musical ability. I. Title.
ML3830.S247 2013
781'.11—dc23
2013003296

British Library Cataloging-in-Publication Data is available

This book has been composed in Sabon

Printed on acid-free paper.

Printed in the United States of America

1 3 5 7 9 10 8 6 4 2

Contents

FOREWORD

When scientific breakthroughs lead to a new level of understanding, some of the scientists involved may, from their improved vantage point, begin to see familiar things in a different light. Music can be one of those familiar things. The tremendous strides made in the nineteenth century in the study of waves and vibration, for example, enabled the great physiologist Hermann von Helmholtz (1821–1894) to see that the ear is constructed like an extremely sensitive frequency analyzer, and that musical consonance and dissonance could be linked to the interactions of the many frequencies registered in the inner ear. Musicians found Helmholtz's research both fascinating and inspiring, and a whole literature of music theory developed around it. The famous physicist Ernst Mach (1838–1916), in thinking about those component frequencies, realized that if two versions of the same tune started on different pitches, one higher, one lower, it would be possible for none of the component frequencies to match. This meant that the two versions were the "same" not by virtue of common sensations (those frequencies) but by virtue of some holistic quality recognized by the mind—a *Gestalt* quality. His fresh insight inspired a generation of psychologists working in the 1920s.

In the second half of the twentieth century, the science of mind was transformed first by a cognitive revolution and then by a revolution in neuroscience. Jay Schulkin, a researcher in behavioral neuroscience, inherited the advances of the first of these revolutions as a student and is today recognized as a distinguished participant in the second. From the fresh perspective of this new science Schulkin looks at music and sees a host of exciting connections between the laboratory and the concert stage, car radio, or iPod. The many aspects of music—its social, emotional, cognitive, somatic, and evaluative components—all have their analog in activities of the human brain. So it makes sense for a neuroscientist, especially one well versed in music, to explain these connections. While Helmholtz detailed the gross anatomy of the ear, Schulkin helps us to understand the fine structure of neurons in the brain and the many special molecules that enable the transmission and modulation of information conveyed from neuron to neuron. These molecules are facts of biology and come with their own traces in the long evolutionary record of life on earth.

Jay Schulkin arrived for graduate study in philosophy at the University of Pennsylvania two years before I arrived there for graduate study in music history and theory. Though we never met as students, our intellectual lives crossed in the person of a storied professor, Leonard B. Meyer (1918–2007). Meyer was a famous humanist, yet one with an abiding interest in science. His curiosity was infectious and knew no disciplinary bounds. When I first entered Meyer's office, issues of the journal *Science* were strewn on the same table with a score of Arnold Schoenberg's adventurous *Pierrot Lunaire*. For Meyer, the love of inquiry and intellectual exploration was something shared by great artists and great scientists alike. He was sure that musical rhythms must be grounded in our physical bodies and their movements, and he was convinced that musical syntax was just another manifestation of how humans interpret the meaning of patterns of every kind. Schulkin is able to show how these speculative ideas from a scholar of music find concrete support in the realm of neuroscience. Meyer would be justly proud.

Two recent books have brought the subject of music cognition to a broad audience. The first of these, *This is Your Brain on Music* by music psychologist Daniel Levitin (2006), provides an engaging tour of the many phenomena involved in the act of listening to music. In the second, *Musicophilia* (2007), the famous clinician Oliver Sacks relates how changes in the brain can dramatically affect a person's musical abilities and enjoyments. Schulkin's *Reflections on the Musical Mind* is for the person who has been intrigued by those discussions and now wants something more, something that can serve as a bridge between the core issues of music and the thousands of scientific studies that each address some particular facet of how our brains engage with meaningful patterns of sound. By taking "an evolutionary perspective" and asking the difficult "why" questions—Why do we have music? Why might it be beneficial for society? Why is music linked with dancing? Why do we respond emotionally to music?—Schulkin engages his reader in issues that have been debated for centuries but that now can be examined afresh.

Robert O. Gjerdingen

Preface

Music has always played an important role in my life. As a young child, the activity I most enjoyed was playing the clarinet. My teacher, Mark Dashinger, taught in the basement of a building in the Bronx called the "Coop," a socialist-oriented remnant from an earlier period. My junior high school music teachers were Mrs. Fragelo and Mr. Masteranglo. The only course in which I received the highest grade, and the only subject that really captured my interest, was the music class.

All three of these teachers assumed I could read music, and I could, a bit, but what I was doing was hearing a piece of music and playing it back almost verbatim. And that was my main talent as a kid. It was a nice talent, but since I had everybody thinking I could read music, it backfired when it came time for tryouts at music schools.

My grandfather spent time with the parents of George and Ira Gershwin, and eventually bought their restaurant from them. I was told on one occasion that George came home from high school and handed his father a sheet of paper with some music on it that he had written. The father did not think much of it and gave it to my grandfather. When he went home, he handed it to my grandmother, who, I am told, threw out the piece. I never talked to my grandfather about the Gershwins, but I grew up playing their music, as well as much of the music of the early and mid-twentieth century. My grandfather, however, loved Brahms.

Amid the turbulence of growing up in the mid-1960s, I spent a lot of time playing music with my two friends, Douglas Mafei and Jeffrey Dix. Improvisation was our aesthetic delight. My two cousins, Freddie and Michael Kaplan, who were Borscht Belt funny and musical with it, worked in a music store in the Bronx; it was always a delight to visit them.

At college my knowledge expanded greatly in the professional ambiance of being surrounded by dancers, musicians, and actors. As an undergraduate, I was especially influenced by the interplay of music and dance in the 1970s and 1980s. To this day, when I hear music I often see movement—not surprisingly, since the brain is a rhythmic generator, and music is tied to our successful movement, as well as to our evolution. For a number of years I lived in a Manhattan apartment building devoted to the performing arts. These experiences only expanded my musical sensibilities.

In graduate school, while still a student in philosophy, I took a seminar with Leonard Meyer, one of the great musicologists of the twentieth century. Meyer had been influenced by the pragmatists (C. S. Peirce, John Dewey, George Herbert Mead), and worked directly with Charles Morris at the University of Chicago. Meyer was oriented to the cognitive revolution of the 1970s, conveying a naturalistic perspective—all music to my ears.

I also interacted during this same period with Peter Marler, an ethologist who was at the University of Pennsylvania for a year and at Rockefeller University. Marler's orientation to animal communication was influenced by C. S. Peirce's writings on semiotics. Most important, Marler understood the cognitive revolution linking it to his studies on song, syntax, hormones, and the brain. He also had a wonderful ability to imitate the diverse species that he studied—again, all music to my ears.

Music, philosophy, and science are continuous—they roll into each other—and that is the way I have experienced them and this is how they are presented in this book. Thus, getting into the rhythm of things, reflecting on music, and then searching for a coherent explanation brings the three together. And I sure have enjoyed this project on music.

For those not cited, I apologize in advance. There are a number of interesting books, some of which I have cited, let alone articles, that overlap with the perspective presented in this book. The field is vast and this book represents just a subset of a rich and beautiful field of inquiry. I want to thank my graduate student Greta Raglan, in particular, for her help on this book. My friends, family, and colleagues, as always, have been very helpful. You know who you are and have been mentioned before. Thank you.

This book is dedicated to the memory of Mrs. Virginia Armat and the many hours we spent singing at the piano in Washington, DC—she was warm and gracious with a wise sensibility, as she ascended to her ninety-second year of a very full life. The book is also dedicated to my modern dance friends from an earlier life, Philip Grosser and Ilana Snyder.

Reflections on the Musical Mind

Introduction

All of us link music to memory, to motivation and to human social contact. Only a handful of individuals may play music, but all can at least sing some, and do so. Music is like breathing—all-pervasive.

Creative expression, such as music, arises from core capabilities within increasing ecological/cultural opportunities.[1] With expanding environments there is a greater diversity of expression. Music is a core human experience and generative processes reflect cognitive capabilities.

Underlying the behavior of what we might call a basic proclivity to sing and to express music are appetitive urges, consummatory expression, drive, and satisfaction.[2] The expression is not as simple as food ingestion, but then that is not so simple either in our species,[3] since the lures encouraging us to eat are boundless.

Music, like food ingestion, is rooted in biology. Appetitive expression is the buildup of a need, and consummatory experiences are its release and reward. Musical appetitive and consummatory experiences are embedded in culturally rich symbols of meaning.

Music is linked to learning, and learning is what we do best; the pedagogical predilection is vast. And, like the consideration of emotions, as Darwin understood, music is a human core capability. Moreover, Darwin did not separate emotions and cognition. I have not either. Emotional systems are forms of adaptation; consider, for instance, the importance for survival of the immediate detection of facial expression. The diverse forms of its bodily expression are emboldened with cognitive capability, rich in information transfer and processing.

Darwin, and certainly John Dewey, understood that emotions are rich in cognitive functions and appraisal processes. There are diverse forms of appraisals and some of these, like music and faces, are affectively opulent. Moreover, the issue is about function. Music is rich in information processing and tied to an appreciation of nature.

Cognitive resources are rich in the generative processes within the expectations that surround music; this puts music within both the sciences and the arts. The cognitive architecture, the generative processes, the diverse variation, and embodiment of human meaning within almost all spheres of human expression are rich fields of discovery for the arts and the sciences.[4] Human meaning is tied to social contact,[5] and music is a fundamental part of human meaning: making it, participating in it, remembering it, and sharing diverse forms of experience in vectors of meaning.

Art, like science, is embedded in discovery, test, experiment, and expansion through technique. There is no divide between the scientific and the artistic. They run into each other quite readily and naturally as they expand the human experience. A romantic sensibility about song, appreciating it as an instrument of music, is a continuous function within the broad array of our investigations. Song either culturally evolves or does not depending upon the resources available and the strokes of genius punctuated within often stable cultures.

Moreover, music is richly filled with emotions. What makes song a good example of emotion is that emotions are bound to the functions of music. Now this might seem a long way from birdsong, and it is, but so are we. For us, music evolved from communicative functions into a highly cognitive action, a piece of aesthetics that happens to occur—and we of course are all the better for it.

Cognitive and Neuroscience Revolutions

I was fortunate to have begun my career within the "cognitive revolution," and now there has emerged a discipline called "social neuroscience." This book is a reflection of both.

By the middle of the 1970s, behaviorism was on the wane and the rigorous methods that emerged as virtues would be dissociated from its ontological claims. All references in behaviorism to the mind are, in some narrow sense, references to behavior without mention of cognition or experience. Behaviorism dominated the field of psychology, like its counterpart psychoanalysis in psychiatry, for much too long. Both had something to contribute, but not to the extent to which they overshadowed their respective fields.

The cognitive revolution is deeply rooted as a reaction to behaviorism, especially via Noam Chomsky's critique of B.F. Skinner in 1957, and in *The Structure of Behavior* by Miller, Pribram, and Galanter.[6] The downfall of behaviorism was that it was utterly inadequate to account for our cephalic capacity. Further, the learning theories that predominated were just too narrow. In this case, Occam's razor cut too close and left a barren landscape. The foundations of sensory data dominated disproportionately. From where we now stand, and in spite of much dissension and a lack of integration among groups, the cognitive revolution is capable of producing an understanding of music.

The cognitive revolution looks somewhat like a resurrection of modern rationalism, but it is broader than that. Concepts are tied to diverse forms of cognitive capacity that inhere in function and adaption, including innate concepts; but that is only part of the story that emerges from the cognitive revolution. Cognitive systems are not antithetical to sensory/motor regions of the brain, along with other regions of the brain vital for perception or action.[7] It is not whether we have concepts that are deeply part of the cognitive architecture versus the sensory pangs of experience; it is just the issue of the adaptation, speed, and coherence of human action and performance.

After all, cognitive systems are embedded in rooted problem solving, something my colleague Mike Power has called "cognitive physiology," and others now call "cognitive biology." To me it is plain old Biology 101: cognitive systems are inherent in biological adaptation. We come prepared with a toolbox of capabilities that allow us to readily recognize animate objects, to sense time and space, to use language, and to discern agency in others.

J. J. Gibson suggested that there is direct cephalic access to environmental sources of information and practices in the organization of action.[8] Thus, some questions are: what are the conditions for adaptation and what are the factors in the environment that allow readily available resources? This view of cognitive resources is linked to the ecological/social milieu, to what is available, what is dependable, what is utilizable, as well as the ability to use and unload information into environments that expand, enable, and bolster memory functions as core cognitive events.[9]

Context helps to facilitate performance, musical, and otherwise. Our ways of hearing and responding to music are steeped in the direct ecological exposure to and expectations about sound and meaning, music and context, pervading our musical sensibilities.[10] A vivid sense of being within a context grounded in forms of ecological validity renders musical experience as palpable a human experience as there is, woven within tapestries of history, invention, and capabilities. It is this sense of grounding that makes features stand out so easily in music and enables the mutualism among the perception, action, and external events that are quite palpable in music.[11] The events are always relative to a framework of understanding—a social context rich in practice, style, and history.

Cognitive expectations are linked to Gibsonian anchoring to events. Core features that enable easy adaptation to the environment are represented in the brain and the social milieu of ritual and performance. Anchoring is one of the features of shared musical experiences, of the rites, rituals, and symbols of our lives pregnant with musical expression,[12] to affordances that are richly endowed with structures that facilitate memory.[13]

The expansion of memory facilitates the wide array of what we do, including music. The emphasis is on action and perception knotted together and coupled to events—in this case musical events - listening to and participating in them.

Table I.1 from Merlin Donald is about memory and action in perception, which crosses virtually the entire realm of musical expression. It is easily adapted to musical sensibilities and capabilities tied to memory coded internally and externally. Internally, it is rather limited, and, externally, it resembles what is sometimes

TABLE I.1

Properties of Internal and External Memory Compared

Internal Memory Record (endgram)	External Memory Record (exogram)
Fixed physiological media	Virtually unlimited physical media
Constrained format, depending on type of record, and cannot be re-formatted	Unconstrained format, and may be re-formatted
Impermanent and easily distorted	May be made much more permanent
Large but limited capacity	Overall capacity unlimited
Limited size of single entries (e.g., names, words, images, narratives)	Single entries may be very large (e.g., novels, encyclopedic reports; legal systems)
Retrieval paths constrained; main cues for recall are proximity, similarity, meaning	Retrieval paths unconstrained; any feature or attribute of the items can be used for recall
Limited perceptual access in audition, virtually none in vision	Unlimited perceptual access, especially in vision
Organization is determined by the modality and manner of initial experience	Spatial structure, temporal juxtaposition may be used as an organizational device
The "working" area of memory is restricted to a few innate systems, like speaking or subvocalizing to oneself, or visual imagination	The "working" area of memory is an external display that can be organized in a rich 3-D spatial environment
Literal retrieval from internal memory achieved with weak activation of perceptual brain areas; precise and literal recall is very rare, often misleading	Retrieval from external memory produces full activation of perceptual brain areas; external activation of memory can actually appear to be clearer and more intense than reality

Source: Adapted from Donald 1991.

called "extended mind."[14] The scope of musical memory is expanded to what Donald calls an "exogram." An exogram is an external record of memory.[15]

All of this contributes to the cognitive neuroscience of different features of musical sensibility.[16] A study of music emphasizes its independence from language while tying it, like all of our cognitive functions, to a diverse set of cognitive capabilities. Moreover, common forms of mental representations underlie action and perception in musical performance and musical sensibility.[17]

Diverse forms of cognitive systems reflect brain evolution,[18] with musical sensibility distributed across a wide array of neural sites, something that Leonard Meyer, an early exponent of a cognitive/biological perspective, appreciated.

Music, Meyer, and Pragmatism

Musicologists like Leonard Meyer understand that uncertainty is a basic fact of our existence, and that cognitive architecture is rooted in determining relevant information, as well as seeing that the search for meaning is adapted to what Meyer in 1967 called "embodied meaning."[19] Embodied meaning or action, a sense of all body kinesthetics, is a popular concept now, very much in vogue. It was commonly understood by such early pragmatists as John Dewey and George Herbert Mead, and also by those they influenced, namely, Leonard Meyer, who studied at both Columbia University and the University of Chicago.

Meyer had a theory of music that was intimate to this sense of action and perception. His is a view of cognitive resources embedded in emotion,[20] in which emotions are tied to action, and embedded in biological problem solving, not perfect problem solving.

What Meyer incorporated into his sense of music was a pragmatism based in Peirce and Dewey's notion of inquiry. Peirce understood that "thinking is a species of the brain and cerebration is a species of the nervous action;"[21] Peirce and Dewey always noted that there are no precognitive events. For both, degrees of cognitive systems underlie perception, attention, and action.

The theory of inquiry, of hypothesis testing, is entrenched in this cognitive perspective; what Peirce called the "fixation of belief" rooted in the organization of action. The orientation is not simply reactive, but anticipatory and also responsive to discrepancy with expectations.

Of course, Meyer and his approach to understanding music was a precursor to the cognitive revolution. He had embraced John Dewey and other early pragmatists with the idea that cognitive systems are inherent in action with anticipatory cognitive systems. Musicologists like Meyer also embraced psychobiology with the idea of something like cognitive physiology. Meyer's collaborations with psychologists (e.g., Burton Rosner, who moved to Oxford from the University of Pennsylvania) were about categorical perception in music.[22]

But these are categories in action. Perception for Meyer, as it was for Dewey, is tied to although not bound by, action. However, movement is not a trivial feature; indeed, now we know that imagining action is adaptive and tied to problem solving. Imagining events is an important feature of adaptive systems. Both Dewey and Meyer were anchored in an evolutionary perspective in which they understood that the animated machinery that permeates musical expression is based in diverse biological adaptations (e.g., anatomy, functions).

In explaining music, at least in part, Meyer noted, "Understanding music is simply a matter of attending to and comprehending tonal-temporal relationships."[23] Underlying the reciprocity of play, regularity, and style is in part the patterns of repetition;[24] and, of course, rhythm is endemic to music.[25]

The cognitive revolution was very much focused on understanding the structure of cognition and the relationships between expectations and subsequent actions. These include musical constraints, like linguistic constraints, for the production of diverse languages.[26] Living within constraints, and also breaking them, are features of the cognitive nature of music; the gaps in our experience underlie the curiosity and the emotions of some of the aesthetic and intellectual exploratory cognitive drives (see chapters 4 and 5).[27]

Music is often functional because it is something that can promote human well-being by facilitating human contact, human meaning, and human imagination of possibilities. We came quite easily, one might surmise, to the cephalic state of enjoying music for itself, its expanding melodic and harmonic features, its endless diverse expression of sound, moving through space, and within our power to self-generate it.

Music, under many conditions, also breaks barriers; Ray Charles, blind since his childhood, was as capable as anyone of singing "Georgia" or "God Bless America." Music has the power to bridge many divides. Perhaps, as Oliver Sacks noted in his book on music, "The image of the blind musician or the blind poet has an almost mythic resonance, as if the gods have given the gifts of music or poetry in compensation for what they have taken away," and thereby provides some "sight" or insight into human bonds that matter.[28]

"Music to my ears" is an old idiom in the English language. It was first attested in James Fenimore Cooper's *Home as Found* (1838),[29] but was probably in popular use well before that. After all, it succinctly expresses in a single phrase a fundamental auditory perception of the way rhythm and music convey pleasure, familiarity, and welcome. As a phrase, "music to my ears" connects the notion of ease with that of aesthetics, a natural property and an event replete with sweetness. While musical meaning is much broader than this, what surely makes music so culturally robust, so much a part of all of us, is the ease with which it manifests itself in our experiences.

Another phrase regarding music that everyone identifies with is "I've got rhythm, who could ask for anything more?" from the George and Ira Gershwin musical *Girl Crazy*.[30] It concisely conveys the way music is linked to movement: tapping, swirling, and dancing. This well-known song can be played, like virtually all music, many different ways and in many different styles. Beginning in 1934, when the Gershwins first played the song publicly, through Benny Goodman in 1943, to Charlie Parker in 1946, and up to the present day, many artists have celebrated the different forms of music in this song. As Lawrence Zbikowski, a musician and musicologist, has well noted in *Conceptualizing Music*,[31] the music of Goodman or Parker maintains coherence, as it is steeped in a set of categories that share a family of characteristics and common themes.

Cognition and Action

Cognition is endemic to musical events and tied to biology, adaptation, and prediction. Indeed, we know that human organization is replete with anticipatory cognitive systems, most of which encompass the vast cognitive unconscious.[32] Action sequences are well orchestrated and embedded in successful survival for both short- and longer-term expression.[33]

One early limitation of the cognitive revolution was its tendency to leave out discussions of emotions: that "other stuff." Too often emotions were thought to render one less rational, less prone to act. After all, the term "passions" traditionally means to be passive, unable to act, and they were thought to render one ineffective. This traditional view dominated the perspective of the cognitivists, and so left out an important part of biology: our emotions and their adaptive nature. Such a view of emotion is now considered extreme and not tied to biological perspective.[34] After all, one can get lost in thought and not be able to act just as easily as one can get lost in emotions and not be able to be effective. Emotions are no less adaptive than abstract thought.

Cognitive expectations knotted to goals generally underpin the organization of human and animal action, as well as music. Cognition is no spectator, and musical expectations are but one example among others. Important biologically derived cognitive systems are not divorced from action or perception, but are endemic to them.[35]

Lakoff and Johnson (1999) depict relationships between perception and action, which underlie all of music, with thinking, perceiving, communicating, imagining, and the like (see table I.2).[36] Music is the kind of thing that is an action but which can also permeate our imagination by imprinting on our neural systems.

Music plays inside our heads, and as we shall see, common neural circuits underlie the action of playing and hearing music, as well as imagining the music in reverberation.[37]

Music is fundamental to humans as a species. Most of the expectations we have may not be explicit, since the vast array of the cognitive systems are not conscious; imagine playing an instrument while being explicitly conscious of all that we have to do. Impossible![38] Cognitive systems are vastly unconscious and underlie action as well as music. The inferences, expectations, and prediction of auditory events are not particularly part of our awareness, and certainly the mechanisms are not.[39]

TABLE I.2

Relationships that Underlie All Aspects of Musical Experience
Thinking (music) as perceiving
Imagining (music) as moving
Knowing (music) as seeing and responding
Attempting insight (through music) as searching
Representing (music) as doing
Becoming aware (of music) as noticing
Communicating (music) as showing
Knowing (music) from a "perspective"
Listening as detecting, knowing

Source: Adapted from Lakoff and Johnson 1999.

Hermann Helmholtz, a German scientist of the nineteenth century, worked on the physiology of harmony and the acoustical representation of sound. He suggested, with regard to music, that "Aesthetics endeavor to find the principle of artistic beauty in its unconscious conformity to law."[40] The vast cognitive unconscious, essential for adaptation to changing landscape, was part of his aesthetic sensibility. Helmholtz merged romantic and Enlightenment neuroscientific interests to understand that perception is mediated by mechanisms of measurement, expectations, and hypotheses.[41]

Imagined Action, or Music and the Brain

Positron Emission Tomography (PET) measures blood flow. It is used as a marker of brain activation. For instance, when subjects were asked to imagine grasping objects,[42] significant activation of regions of the brain concerned with movement occurred. In further studies that used neuromagnetic methods to measure cortical activity, the primary motor cortex is active both when subjects observed simple movements and when the subjects performed them.[43]

Of course, the motor cortex is activated in a wide array of human cognitive/motor activities. Importantly, motor imagery is replete with cognitive structure and is reflected in the activation of neural circuitry.[44] So too is auditory imagery reflected in different regions of the brain, including anticipatory musical imagery.[45]

In another study focusing specifically on sensory events in a functional magnetic resonance imaging scanner (fMRI), subjects were presented with spoken words via headphones. Then, in a second part of the experiment, the same individuals were asked to identify the words with silent lip-reading.[46] Not surprisingly, many of the same cortical regions were activated.

In other words, hearing sounds is like imagining them. Both tasks recruit many of the same brain regions.[47] Across a number of perceptual experiences, imagining employs the neural systems involved in seeing them, hearing them, or touching them. Imagine a visual rotation, for example, versus actually looking at a rotating object; it takes a similar amount of time.[48] Moreover, very similar neural circuits are also activated when the object is imagined or viewed.[49] Imagining is the process of creating brain stimulation internally that is similar to what would be created by external stimulation. In other words, the neural structures that are active in imagining objects appear similar to those structures that are active when looking at the objects.[50]

Not surprisingly, hearing music activates many of the regions linked to auditory perception. However, regions of the auditory cortex are also activated when subjects are asked to imagine music or other auditory stimuli (figure I.1).[51]

Thus, despite the difficulty in the end of not knowing what people are actually imagining, one can dissociate hearing something from seeing it through diverse regions of the brain. Of course, the inverted spectrum problem (whether you and I really see or hear the same thing) is always humbling and reminds us that it is difficult to discern what others see and hear, as well as the corresponding relationships to what we report and observe behaviorally.

Perhaps one is now in a better position to understand the genius of Beethoven; deaf for years, he must have heard music imaginatively to compose the way he did. Think of the cognitive

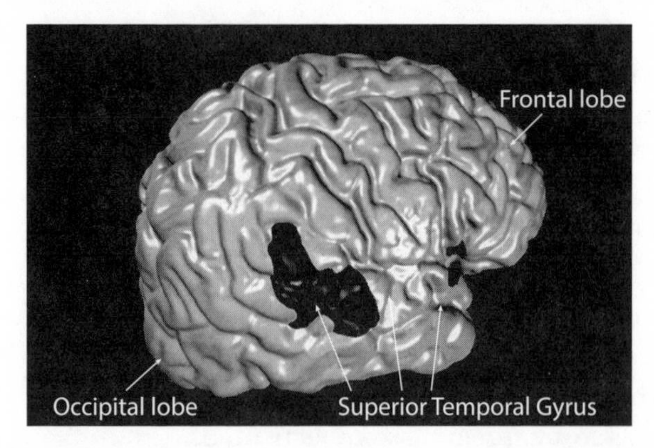

Figure I.1 A neuroimaging scan revealing that even in silence the auditory cortex, pictured here in the posterior portion of the right superior temporal gyrus, is activated.

Source: Reprinted from *Neuron*, vol. 47, Zatorre and Halpern, "Mental Concerts," 2005, with permission from Elsevier.

complexity, the richness of the later parts of Beethoven's life. In fact, we now know that musical hallucinations are often a feature of acquired deafness such as Beethoven's.[52]

Beethoven also reveals a spiritual preoccupation as he probes the depth of human expression;[53] the Ninth Symphony represented an age of romantic possibilities with such freedom of expression by one whom some have considered the most creative genius of his historical period—those last quartets are simply mind-boggling to musical sensibility.[54]

Of course, it also becomes somewhat easier to understand that the same "music to one's ears" may not be heard by one's neighbor. Beethoven is one thing, the rest of us quite another. Yet, the recruitment of cortical regions is generic.

Embodied Cognition

We are now in a better position to understand the neurobiology of musical sensibilities. The neuroscientific revolution of the 1970s and 1980s provided a better understanding of anatomical connectivity. Through anatomical study we now know that the cortical structures are in direct contact with the viscera;[55] the contact is one synapse or connection away from the peripheral guts, the brain stem, and the cortex. Music warms and cools the brain quite directly and rapidly.

The emergence of the cognitive sciences has allowed us to understand the dominance of information systems in the brain, and as such we can now appreciate the diverse information processing systems and appraisal systems that are ingrained in the brain. The cognitive systems are larger than the class of syntax narrowly defined in terms of language, but semantic systems underlie diverse semiotics of song, language, movement, and the organization of action.

Another recent revolution in science is in understanding the diverse information molecules that inform the brain, relating peripheral gut reactions to central orchestration of behavior. The list is quite long and some items are very familiar: steroids, peptides (e.g., oxytocin, vasopressin), and neurotransmitters such as dopamine. Dopamine is an adrenal hormone synthesized in a part of the medullar region. Dopamine is also produced in the brain. Catecholamines like dopamine are tied to action, prediction of events, reward, syntax, and musical sensibility (see chapters 3 and 4, and figure I.2).

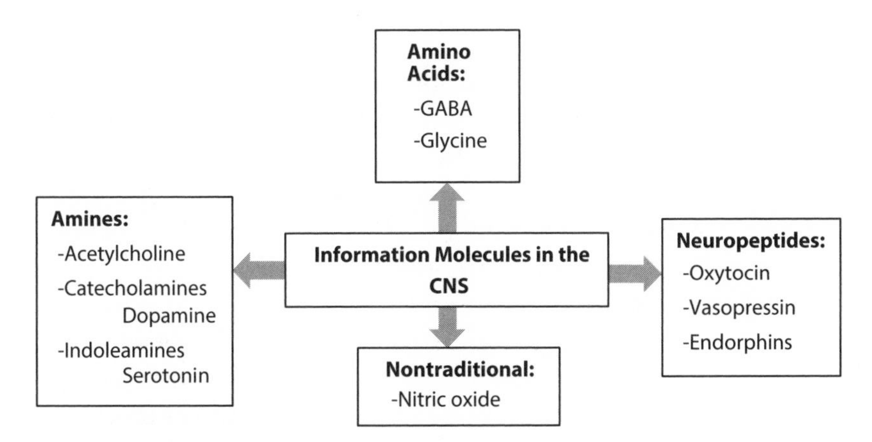

Figure I.2 Examples of important information molecules found in the central nervous system.

Neurotransmitters tend to be broad in terms of regulation.[56] We typically associate dopamine with the diminished regulation and expression of Parkinson's disease, but dopamine also underlies all aspects of the organization of action.[57] Hormonal messengers are a fundamental information class of molecules, of which there are many besides dopamine and norepinephrine. Central norepinephrine is tied to attention.[58] They are the same molecules involved in the organization of behavior as physiological signaling systems.[59]

Information molecules have a phylogenetic history of evolutionary recruitment into diverse functions. The techniques of molecular biology, regulatory physiology, and behavioral neuroscience anchor them to function. Function and adaptation are what underpin biology. Song has long been viewed in the context of function, as it still is in birds. However, song also expands with cultural legacies and expression. Most people's enjoyment of Ray Charles or Beethoven serves no specific function. Biology is one thing; overzealous adaptationism is another. My hope is that this book reflects the former and not the latter.

Music and Culture

Music is a binding factor in our social milieu; it is a feature with and about us, a universal still shrouded in endless mysteries. How music came into being is, like most other features in our evolution, hard to pinpoint. Evolutionary evidence over a wide range of cultural groups reveals diversity of song and instrument, yet gaps and speculative considerations remain: some cultures sing a lot, some sing less, but most do sing.[60] Music is typically something shared, something social; we may sing in the shower or on a solitary walk, but music is most of the time social, communicative, something expressive and oriented toward others.[61]

Music cuts across diverse cognitive capabilities and resources, including numeracy, language, and space perception. In the same way, music intersects with cultural boundaries, facilitating our "social self" by linking our shared experiences and intentions. Perhaps one primordial influence is the social interaction of parental attachments, which are fundamental to gaining a foothold in the social milieu, learning, and even surviving; music and song are conduits for forging links across barriers, for making contact with others, and for being indoctrinated within the social milieu.

Ian Cross and Iain Morley[62] have pointed out the floating, fluid expression of music. There is little doubt that the fundamental link that music provides for us is about emotion and communicative expression, in which the prediction of events is tied to diverse appraisal systems expressed in music.[63]

Music is fundamental to our social roots.[64] Coordinated rituals allow us to resonate with others in chorus,[65] for which shared intentional movements and actions are bound to one another. Perhaps not surprisingly, a philosopher of the Enlightenment would glorify music as a state of nature, putting it squarely within a civilizing factor in some fantasized natural inclination of well-being. Jean-Jacques Rousseau, in fact, had considerable musical talent, writing and winning awards for his music early in his career. Operatic songs were a feature of his aesthetics, with song and sensual experience permeating his early life, a life that was also filled with strife.

Rousseau was fortunate enough to be steeped in song, in music—a glorification of the state of nature. Indeed, the French Enlightenment was rich with theories regarding the origins of music and language (e.g., D'Alembert, Condillac, Diderot, Rousseau),[66] which would come to dominate the landscape as "the language of music," featuring a musical vocabulary connected to diverse emotions.[67]

For Rousseau, Homeric tone and expression suggested an important link between verse and song. In poetry and music, melodic tones for which "the beauty of sounds is natural"[68] are pervasive, while suggestions and responsiveness to the muses appear in what Nietzsche later identified in *The Birth of Tragedy* as music with the Dionysiac urge of self-expression and freedom.[69] Music is a primordial wellspring of utter human creativity and expression, in which music and the muses are at the forefront, breaking boundaries that divide us as a feature of the human will,[70] tied to cephalic capabilities, and reflecting social context and human meaning.[71]

Social meaning in music is particularly clear during the subsequent Romantic period. It pervaded nineteenth-century Europe;

Hector Berlioz is a good example, with his symphonic fantasies that breed human possibilities of hope and transformation.[72]

Culture-bound music is a shared resource that is tied to diverse actions, including sexual function. Yet, sexual function[73] remains but a small piece of where our species has taken music.

Music permeates the way in which we coordinate with one another in rhythmic patterns, reflecting self-generative cephalic expression[74] attached to a rich sense of diverse musical semiotics and rhythms.[75] Music is embedded in the rhythmic patterns[76] of all traditional societies. Our repertoire of expression has incurred a crucial advantage: the ability to reach others and to communicate affectively laden messages.

The social communicative bonding of the wolf chorus is one example that comes to mind,[77] a great chorus of rhythmic sounds in a social setting. A common theme noted by many inquirers is the social synchrony of musical sensibility.[78] The motor sense is tied directly to the sounds, synchrony, and movement. Sometimes the actual motor side of singing is underappreciated.[79] Neurotransmitters, which are vital for movement, are tethered to syntax and perhaps to sound production (see chapter 3). The communicative social affective bonding is just that: affective. This draws us together and, insofar as we are a social species, remains essential to us; a chorus of expression in being with others, that fundamental feature of our life and of our evolutionary ascent. Music is indeed, as Timothy Blanning noted, a grand "triumph" of the human condition, spanning across cultures to reach the greatest of heights in the pantheon of human expression, communication, and well-being. It is in everything.[80]

A natural inclination to observe wondrous nature, to paraphrase Immanuel Kant,[81] is to capture something aesthetic and awe-inspiring. Music is a piece of nature, our nature. Nature allows us to be close to a sublime essence: to be as serene as a fresh dawn and dusk, or as fearful as a threatening bolt of lightning near one's head. Dewey similarly emphasized that art as experience is pervasive.[82] This is a book rooted in social and behavioral neuroscience. Its contents make contact with diverse disciplines. They range from evolution and comparative biology, to musicological considerations about the origins of music. The book is centered in behavioral neuroscience with observations that are both specific to music and more general to basic brain function. Music serves many functions in our lives. From the first sounds to some of the last, human well-being and music are extant across the life cycle of human experience.

In what follows, I give a sense of the origins of music, from the appearance of the relevant anatomical features, to the development of diverse forms of biological systems that figure in musical expression (chapters 1–3). Music reflects our social nature and is tied to other instrumental expression in the adaptation to changing circumstances. Indeed, expectancy and violations of those expectations in music linked to memory and human development (chapters 4–6) are critical features in the aesthetics of musical sensibility (like other avenues of human experience). Music is also connected to movement and dance (chapter 7). The same cognitive factors that underlie what we expect, and variation on the expected, inhere in musical experiences. Moreover, diverse information molecules (e.g., dopamine, and peptides such as vasopressin/vascotocin or oxytocin) that may be involved in the organization of musical expression are imbedded in core neural circuits of the brain that underlie a wide array of human experience. Without music, our world is endlessly impoverished.

The book does not present a grand synthesis or one theory about music; what it does is bring together diverse fields of inquiry that matter in understanding music and human experience. What is special about music? Music contributes to the vast array of human social solidarity and individual expression amidst creative intensification of human experience, within an endless auditory expansion of human sensibility and general well-being.

CHAPTER 1

Music and the Brain

An Evolutionary Context

We are a species bound by evolution and diverse forms of change, both symbolic and social. Language and music are as much a part of our evolutionary development as the tool making and the cognitive skills that we traditionally focus on when we think about evolution.

As social animals, we are oriented toward sundry expressions of our conspecifics that ground us in the social world,[1] a world of acceptance and rejection, approach and avoidance, that features objects rich with significance and meaning.[2] Music inherently procures the detection of intention and emotion, as well as whether to approach or avoid.[3]

Social behavior is a premium cognitive adaptation, reaching greater depths in humans than in any other species. The orientation of the human child, for example, to a physical domain of objects, can appear quite similar in the performance of some tasks to the common chimpanzee or orangutan in the first few years of development.[4] The understanding of objects in space, quantities, or drawing inferences is not that far apart. This is not so for problems requiring a vast array of social knowledge. What becomes quite evident early on in ontogeny is the vastness of the social world in which the human neonate is trying to gain a foothold for action.[5] Music is social in nature; we inherently feel the social value of reaching others through music or by moving others in song across the broad social milieu.

In this chapter, I discuss how music fits into the evolution of our cognitive capabilities, and how the auditory system, larynx, motor systems, and cephalic expansion underlie the expression of music and the evolution of social contact.

Cognitive Capabilities and Problem Solving

Theodosius Dobzhansky is often cited for his remarks regarding how all things are linked to evolution.[6] A biological perspective is the cornerstone in understanding our capabilities, with our musical ability being just one of these, including our sense of space and time, our ability to assess probabilities (the prediction of events), our numerosity, and, of course, our language abilities.

The specific adaptation for decoding facial responses, and the more general aptitudes such as applying numerical capabilities to diverse problems, pervade a biological understanding of cognitive adaptation. Cognitive systems run the gamut across the nervous system. Cognition is not simply defined by a province of the neocortical tissue, the most evolved tissue; cognitive systems are distributed across neural systems that traverse the brain stem to the forebrain.[7] As I have indicated, and will continue to note throughout this book, regions of the brain that underlie musical sensibility and expression are also widely distributed.[8]

Cognition rests at the heart of human understanding. Table 1.1, originally created by the evolutionary anthropologist Steven Mithen, highlights some of the core features of problem solving and human expression.[9]

TABLE 1.1

Forms of Problem Solving and Human Expression

Language Capability	General Intelligence	Social Capability	Natural history Recognition	Technical Knowledge

Source: Adapted from Mithen, 1996.

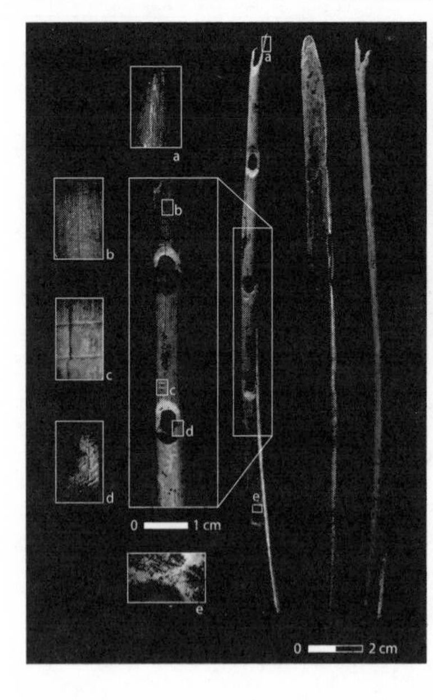

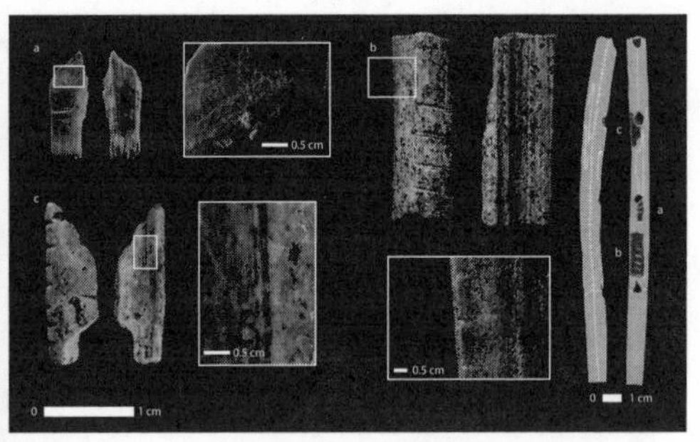

Figure 1.1 Bone and ivory flute fragments from the Hohle Fels and Vogelherd caves in southwestern Germany.

Source: Reprinted by permission from Macmillan Publishers Ltd: *Nature*, Conrad Malina and Munzel, "New flutes document the earliest musical tradition in southwestern Germany," copyright 2009.

These are fairly diverse sets of cognitive predilections that underlie our evolution and figure into much of what we do. Cave painting, tool making, and other skills, as well as our sense of enjoyment in what we do, are embedded in cognitive adaptation and a search to understand something about our surroundings: what to expect, how to cope, how to transform, and so on. After all, *Art As Experience*, as Dewey understood, is in understanding, building, and representing affective content.[10]

From simple tools to facile musical instruments, to elementary symbols, all of these represent a small leap for humankind. For example, see the flutes depicted in figure 1.1. Diverse forms of art and probably music emerged in early *Homo sapiens*, and are evident in remains that date back to at least 40,000 years ago.[11]

Knowledge and a sense of aesthetics are entwined. They coalesce as heightened appraisals predominate in functional contexts; then reprieves occur, long silences and meditative calms amidst the hustle and bustle of life. Reconsidering and musing take precedence amid ephemeral moments of reflection and meditation; artifacts take shape and expand from narrow confines to extended attraction from normal bodily expression. The body is expanded through meditation with a direction set in motion by an evolving brain.

One cognitive adaptation is the capacity for the basic discernment of inanimate objects from animate objects. We have adapted this fundamental cognitive perception into a source of music, art, and religion. We represent animate objects, often giving them divine-like status, which infuses them with specific and transcendental meaning. This is part of a basic adaptation to discern useful objects.

Musical instruments ultimately derive from this expanded cognitive approach to objects. A key artifact is something that is sometimes called a "sound tool" or "lithophone." The oldest date back some 40,000 years ago from sites in Europe, Asia, and Africa.[12] Sound tools are simple stones that resonate when struck, as shown figure 1.2.

Figure 1.2 Flint sound tool, known as a lithophone.

Source: Rock Harmonicon, by William Till. Ca. 1880. United Kingdom. Gneiss and hornblende schist, H. 106.7 × L. 247.7 cm (42 × 97 ½ in.); Longest stone: 77 cm (26 in); Shortest: 22.9 cm (9 in.). The Crosby Brown Collection of Musical Instruments, 1889 (89.4.2931). The Metropolitan Museum of Art, New York, NY, U.S.A. Image copyright © The Metropolitan Museum of Art. Image source: Art Resources, NY

While most of music is song, and song preceded musical instruments, the cognitive/motor cephalic capability for the invention of tools is embedded in music and meaning, with social contact inherent in these events.[13] After all, making objects, musical and otherwise, is a cephalic extension of the world beyond ourselves. The terrain changes, and we scaffold with the broadening array of musical meaning.[14]

Representing objects, dividing kinds by naming and tracking them,[15] is also fundamental to human evolution. Cognitive capacities continued to expand as we explored new terrains and survived in them. Thus, representation is not something that removes us from objects. Instead, the cognitive expansion into art and the knowing process of reflection and meditation provides ways of coming to understand the objects that matter and mean something to us.

Time and Timing

Time is not a thing. Despite the fact that we know something about the neural systems that regulate the perception and organization of time, this concept is still highly theoretical; it is a cardinal feature of our cognitive apparatus, with phylogenic roots in basic clock mechanisms.[16] We have used our diverse cognitive resources to expand upon our grasp of time; we recognize simple concepts of periodicity and precise timing, as well as the timing of other events and, of course, the coordination of events.

Counting, numerosity, and keeping track of events are inherently related. Assigning numbers to objects inheres everywhere and in everything; numerosity may even be shared with other species, since it is so basic to cognitive capabilities. Numeration is also important in reasoning. Demythologized, reason is what we mean by cognitive systems: problem solving. Numbers are linked to the timing of events, staging events in time, and tracking them into coherence and adaptation.[17] These are core capacities that underlie musical expression: the timing of events, keeping track of rhythmic impulses built into the hardware of the brain. The manipulation of numbers in the abstract and in concrete space pervades our cognitive architecture. It is also inherent in our musical expression.

Calculating

Calculating, trigonometry, algebra, calculus, and diverse forms of geometry are embedded in capabilities for keeping track of events This much is true of the Leibnitzian conception of the vast array of mathematic capabilities in which core notational systems that have coherence and symbolic/logical/numerical expression predominate in us. The brain is not simply a calculating machine, but numerosity certainly remains in our attempts at coherence, with calculations permeating every part of our lives.[18]

Some individuals are better than others at framing events in numerosity. Most people fall within a middle range, with the extreme curves representing the best and the worst. Musical capabilities can be graded along these same curves. As far as we know, however, this curve of abilities never matches exactly with numerosity. There is no one simple isomorphic relationship between math and music, or between a feel for numbers and a sense for time, at least not one that can be expressed explicitly. Cognitive capabilities are for the most part unconscious.[19]

Irving Berlin had little schooling, was unable to read music, and had minimal skills in musical virtuosity. Yet, he could write songs that came to define America in the early twentieth century. From "God Bless America" to "The Easter Parade," his were the songs that linked vaudeville to the shining stages of Broadway. "Two Finger Irving" could belt out these themes through simple songs. After all, song captures both the vast complexity of our worlds and its refined simplicity.[20]

Perhaps Berlin was like the classic depiction of the slave who knows geometry, as depicted by Plato in the *Meno*. When asked the right questions, he was quite capable of expressing complex geometrical knowledge. Ultimately, spatial computation is fundamental, and we know that it is expressed in all species studied.[21]

Evolution

The human mind and its capability, including geometric ability and numerosity, and more generally a capacity to grasp a wide array of concepts and their instantiation across diverse sensory expressions, are outstanding. The theory of evolution, a supposition profound in its capture of links to nature, to adaptation, and to function, is not as direct as one would expect. Natural selection, of which sexual selection and sexual dimorphism are derivatives, is not narrow and one-dimensional; it does not advocate reducing all human expression to narrow biologicism or simple adaptationism.

Surely, not all of human behavior is an adaptation. However, core abilities are rooted in biological capability. Small changes grow into larger changes as speciation and habitat take precedence over shape, morphology, cephalic expression, and capability. Wondrous nature was described in the pen and eye of portrait painters with a realistic lens for nature.

Darwin was prepared to believe that musical expression, as a particular universal human expression, is a feature of natural selection, linked to communicative function and sexual selection.[22] Perhaps it is tentatively tied in origins to basic functions, but surely one wants to be respectful of these simple origins without being reduced to them.

Evolutionary trends are not necessarily consistent, as Darwin had suggested and had penned in one of his rather unaesthetic drawings. Evolutionary trends may be more like fits and starts, punctuated by sudden changes.[23]

Some of the key events and core features of our evolution over the last 300,000 years are depicted in table 1.2, in terms of key discoveries of modern human evolution.[24]

One view of evolution is the hypothesis that language and speech emerged some 50,000–100,000 years ago, and artistic representation can be traced back to 30,000–40,000 years ago.[25]

Amidst some presumably unusual and unstable environments, a cognitive expansion, or a change in brain development, facilitated tool use and social cooperative behaviors that proved fateful for our development.[26] Table 1.3 outlines some cognitive cultural features in diverse *Homo* species related to a common modern ancestor; social features, essential for group formation, are pervasive.

Again, we know now that diverse forms of hominids competed and perhaps interbred during the same time period. *Homo sapiens* eventually came to dominate the landscape as other human-like

TABLE 1.2

Primary Features of Human Evolution Over the Last 300,000 Years

280,000 years ago	Stone blades and other pigments are used by ancestors of *Homo sapiens* in Africa
195,000 years ago	Fossil evidence of *Homo sapiens* in Ethiopia
140,000 years ago	Evidence of long-distance exchange and shell fishing in Africa
130,000 years ago	*Homo sapiens* move into the Levant
80,000 years ago	*Homo sapien* occupation of the Levant ends; replaced by the Neanderthals
77,000 years ago	Ornamental shells and geometrical engravings made in South Africa
50,000 years ago	*Homo sapiens* expand out of Africa
41,000 years ago	Oldest evidence of humans in Europe
36,000 years ago	Oldest European cave art
33,000 years ago	Neanderthal-crafted ornaments suggest they imitated *Homo sapiens*
28,000 years ago	Neanderthals become extinct
27,000–53,000 years ago	Estimated age of the youngest *Homo erectus* fossils in Indonesia
20,000 years ago	*Homo sapiens* move from Asia to North America
18,000 years ago	Youngest known fossils of *Homo floresiensis*

Source: Mellars 2006b; Zimmer 2005.

primates became extinct.[27] Some of these hominids, including our own direct ancestors, must have expressed musical sensibilities at some point.[28]

The literature is brimming with a vast array of theories regarding brain size and evolution. Depicted in table 1.4 is a summary of diverse studies regarding the relationship between neural weight and evolution. We can discern that perhaps from *Homo erectus* to *Homo sapiens*, music may have been a feature of our own primate line based on cephalic weight. See table 1.4 for a comparison

TABLE 1.3

Hypothesized 'Cultural' Properties of Hominin Taxa

	Homo ergaster	*Homo neanderthalensis*	*Homo sapiens*
'Cultural' Inference	Greater planning depth	Greater planning depth	Strong ethnicity
	Imitation	Ethnic affiliation	Extensive symbolism
	Limited innovation	Symbolic thought	
	Emotional affiliation?	Language	
	Theory of mind?		
Observation	Significant meat-eating	Flexible technology	Local networks
	Delayed growth	Regionalization	Cultural replacements
	Complex tool making	Music	Rapid change
	Persistent traditions		Material diffusion
	Music?		Music

Source: Foley 2006.

across species, that lived either millions of years ago (mya) or thousands of years ago (kya), of respective encephalization quotients (EQ). Encephalization quotients are used to measure the expected cognitive capabilities of a species, using the ratio E_a/E_e, where E_a represents the brain mass of the given species, and E_e is the expected brain size for that species based on a "standard" species of the same taxon. EQ represents the deviation in actual brain size from the expected brain size for that species.[29]

Figure 1.3 depicts an endocast of the frontal region of a putative *Homo* around 2 million years ago, compared to the frontal planes of (a) chimpanzee, (b) orangutan, (c) gorilla, and (d) modern human.[30] Additional neocortex, particularly that of the auditory

TABLE 1.4

Comparison of Encephalization Quotients Across Primate Species

Species	Date Range	EQ
Australopithecus afarensis	3.9–3.0 mya	2.5
Paranthropus boisei	2.3–1.4 mya	2.7
Paranthropus robustus	1.9–1.4 mya	3.0
Homo habilis	1.9–1.6 mya	3.6
Homo ergaster	1.9–1.7 mya	3.3
Homo erectus	1.8 mya–200 kya	3.61
Homo heidelbergensis	700–250 kya	5.26
Homo neanderthalensis	250–300 kya	5.5
Homo sapiens	100 kya–present	5.8

Source: Power and Schulkin 2009.

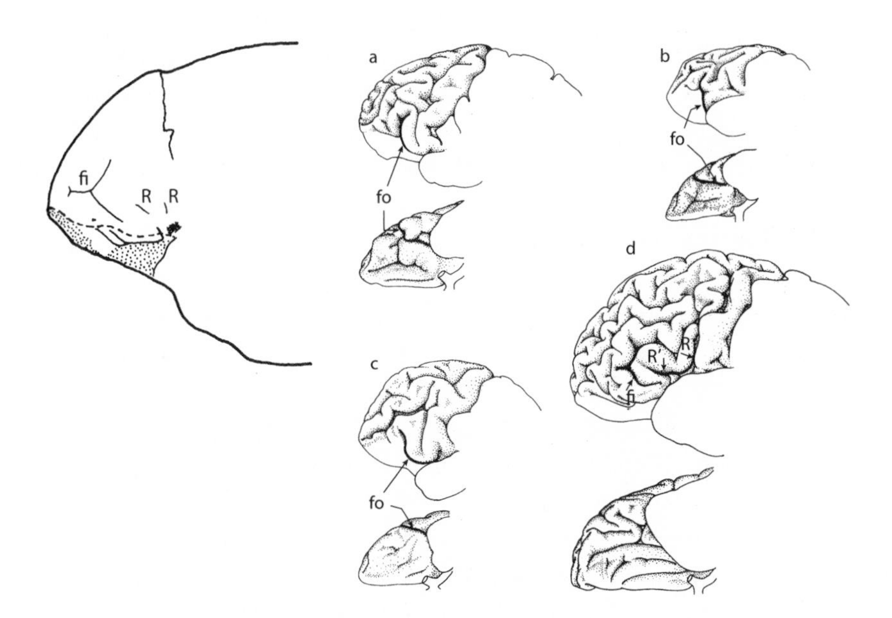

Figure 1.3 Brain evolution across diverse species.

Source: From *Science*, 221: 4615, Sept. 9, 1983, Falk and Dean, "Cerebral Cortices of East African Early Hominids." Reprinted with permission from AAAS

cortex, facilitated the conditions for greater social/communicative/instrumental functions, including that of music.

In the process of evolution, what exploded in our species was the visual system. Commanding the greater part of the neocortex, this expanded visual system reflects our dependence upon sight. It allows us to coordinate with others in joint social behaviors and to warn against constant danger.

The degree of cortical expansion is directly correlated with vision. This is most clearly seen when considering the expansion of vision in comparison to non-vision.[31] Vision is critical to keeping track of and working with others.

But it is hearing, not vision, that is fundamental for music, and so we turn our attention to audition.

Hearing and Vocal Expression

Auditory sensibility and anatomy underlie hearing and vocalization. Harmony, being an evolving feature, is bound to pitch perception and generation as a fundamental trait.[32] The auditory system, like all sensory systems, traverses the whole of the brain, from the peripheral cochlea to the auditory cortex as depicted in figure 1.4.[33] As we shall see, musical training impacts the auditory system, in addition to other cortical regions, in diverse ways.[34]

During musical training, various regions of the brain are recruited across the range of auditory cephalic competence for timing, pitch, and timbre. The cephalic system comes prepared to learn about music,[35] just as it comes prepared to learn about shapes, tastes, and objects.[36]

Language and tone of voice are early social bonds. In combination with other sensory systems pregnant with cephalic function, they form social contact for the neonate in the larger social world. Tones and song have long been noted to be important adaptive factors in the facilitation of early social attachment. A sense of tonal structure may be stimulated by exposure to music during early development, in which expectations for tone and musical structure emerge.[37]

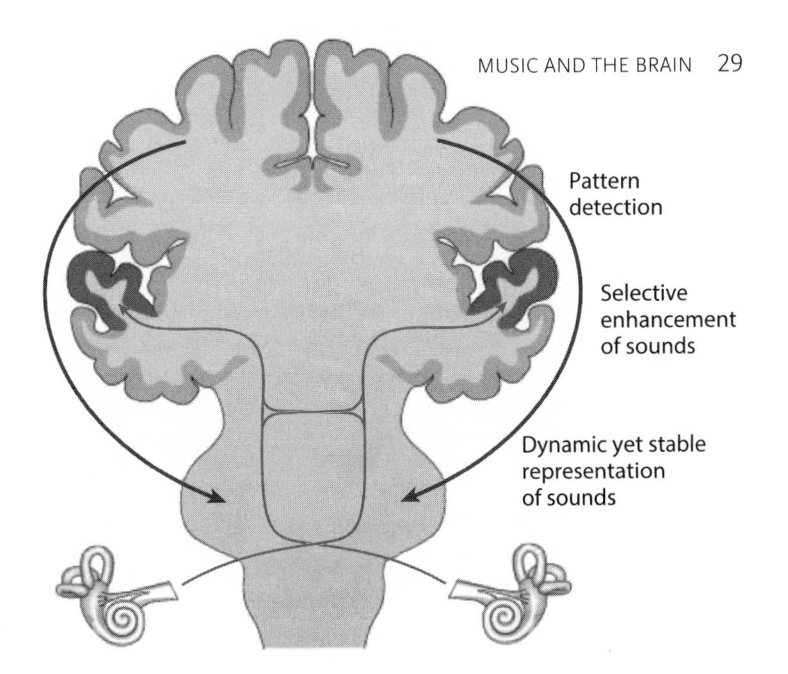

Figure 1.4 Illustration of the interplay of diverse systems in musicians.
Source: Kraus and Chandrasekaran 2010.

Hearing as Seeing, Seeing Through Hearing

I remarked earlier that we often associate heightened musicality with blindness. This may be more than a mythological perception. My colleague at Georgetown, Josef Rauschecker, has suggested that Hebbian "synaptic efficacy" is clearly operative in blind individuals.[38] In other words, the brain compensates for the loss in one system by the expansion of another, a common theme perhaps throughout cephalic adaptation. Regions of the occipital lobe are activated by sounds in blind individuals. Indeed, sighted individuals deprived of visual stimuli, even after a few days, begin to express greater activation to non-visual stimuli in the occipital lobe (e.g., audition and touch). What is interesting to note, and in concert with enhanced cephalic compensation, is the finding that absolute pitch competence is greater in blind musicians.[39] Regions of the superior temporal lobe seem to integrate both auditory and visual information.[40] Sounds and songs are the semiotic vehicle for social contact, social alliances, and social avoidance, the sources of acoustically rich fields of information.[41]

It is also noteworthy that even children who are blind from birth still understand something about spatial objects and concepts.[42] The sensory systems are replete with cognitive resources, emboldening the organization of action vital for the formation of social contact and social aversion.

Importantly, in congenitally blind individuals, there is an expanded auditory cortical system (and no doubt other sensory systems) to compensate for the decrease in sight.[43] In blind subjects, regions of the occipital region, which are activated by only visual stimuli in sighted persons, are activated by audition and thereby expanded to auditory function.[44]

Core Motor Systems: The Larynx and Cognitive Motor Expression

What produces song is a core anatomy that includes a larynx of a certain shape and perhaps size, tied to systems that orchestrate movement featuring statistically related acoustical harmonics and periodicity.[45] These are bound to preferences for ratios and intervals between sounds via the modulation of the larynx.

The expansion of the larynx, and the development of cognitive/motor capability, underlies speech, song, and other social communicative cephalic expressions (see figure 1.5). These features figure in key adaptive responses that underlie our social capability.

Thus, the opportunistic extension of diverse systems may have played an important role in song, and perhaps syntax, and ultimately music. Access to pre-adaptive systems makes a difference in diversity of expression.[46] In an interesting analogy, Daniel Lieberman, an evolutionary biologist, put it this way when reflecting on the larynx: "The larynx is a source of acoustic energy, not unlike the reed in a wind instrument."[47] Communicative capabilities are endlessly opportunistic in the exploitation of existing resources with diverse and expanding uses. Sound production itself is well understood and involves the larynx and the vocal cord or tract.[48]

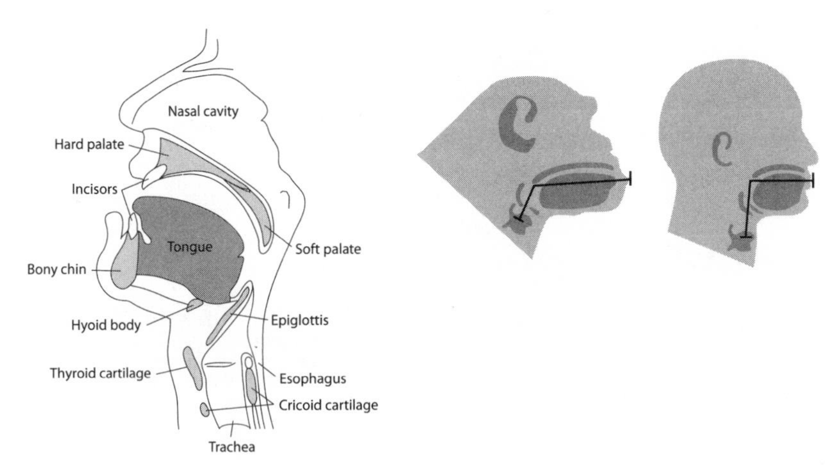

Figure 1.5 Key features in the vocal capability of a chimpanzee (center) versus a human (left, right).

Source: Lieberman and McCarthy 2007.

More generally, auditory perceptual systems code and structure events for music within contexts of semiotic systems, which then further expand our capabilities for song. It probably started in small steps; an evolving motor cortex united with cognition and perception underpins the production and appreciation of song.[49]

"Singing Neanderthals"

Some 30,000 years ago, evidence suggests that Neanderthals and *Homo sapiens* co-inhabited overlapping geographical locations, in which climate change was a significant factor in the Neanderthals' demise.[50]

Figure 1.6 charts the divergent paths of Neanderthal and *Homo sapiens*. The first Neanderthal remains were found in 1856 in Germany's Neander valley.[51] Subsequent discoveries indicate that the species roamed over a large area of Europe, the Middle East, and Central Asia from the period of the last Ice Age, well before *Homo sapiens* moved out of Africa. Changing weather conditions may have set the stage for the demise of the Neanderthals and the

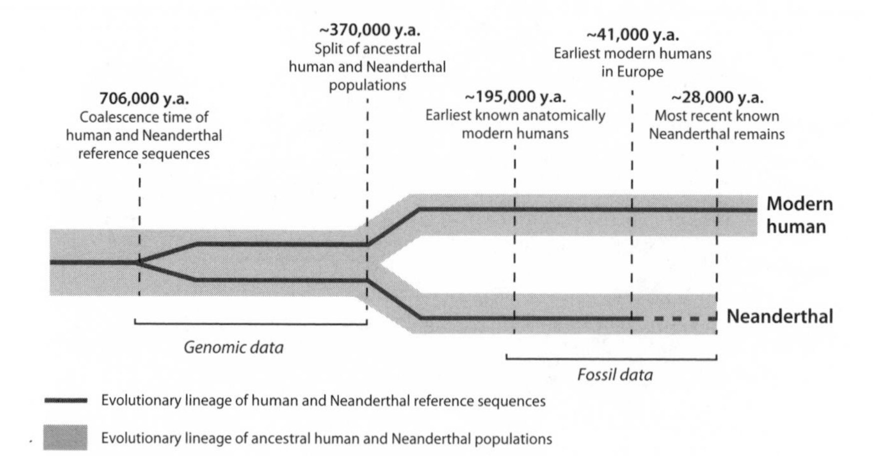

Figure 1.6 Illustration of divergent genomic data for Neanderthals, humans, and ancestral humans, relative to critical events across evolution.

Source: Noonan et al. 2006.

sustained capability of *Homo sapiens*. Paleoanthropologists view our broader cephalic capacity for adapting to all kinds of change as one possible reason for our survival while Neanderthals became extinct.

A number of archaeological sites have found a common overlap of Neanderthal and human remains.[52] Some academics now believe *Homo sapiens* and Neanderthals may have interbred. Recent developments in genetic testing of fossil remnants have provided us with some insight into the Neanderthal genetic makeup that may eventually help us to trace the lineage and extinction of the Neanderthal. Any transfer of genetic information between modern humans and Neanderthals probably occurred during encounters between 20,000 and 30,000 years ago. This means that modern humans outside of Africa may have up to 2 percent of inherited genes traceable to Neanderthals.[53]

Steven Mithen, a biological archaeologist, suggests that the Neanderthals were a remarkable example of the *Homo* line, perhaps with a gift for song, but less facility for speech and language. Given the possibility of interbreeding, it is clear that on the genetic level, the differences between Neanderthals and *Homo sapiens* are quite small.[54]

A gene linked to speech and language (FOXP2) is found in both Neanderthals and *Homo sapiens*.[55] The FOXP2 gene, tellingly, is implicated in speech disorders and language deficits.[56] It is also known to disrupt the neural circuitry that regulates the cortico-basal ganglia.[57] Human cognitive systems for speech lie in regions of the brain including the basal ganglia and frontal cortex. We know that small differences at the genetic level can result in dramatic changes in form and function.[58]

While both species were well endowed with cognitive and symbolic capacities, Mithen suggests that *Homo sapiens* emphasize language and its underlying syntax, while the Neanderthals stressed singing.[59] *Homo sapiens* have much greater communicative and generative processes that enhanced diverse forms of cognitive adaptation and expansion. Song and dance could only go so far; language is ultimately a more flexible tool, as it underlies many of our core capabilities as a species.

However, the Neanderthals' brain morphology and size, evidence of social groupings, and complex behaviors such as burial practice, suggest high cognitive function with diverse forms of representative abilities,[60] perhaps including an elaborate form of song production without the expansion of linguistic competence.[61]

Mithen describes the proto-musical expansion in which an evolving communicative skill, he speculates, predominated. The suggestion is that Neanderthals' communication was more holistic than the language of *Homo sapiens*, which is primarily information-bearing.[62] Both forms of hominid shared a complex molecular biology with significant cognitive and social capacity, and of course, the fundamental shift that occurred millions of years earlier: bipedalism.

Of course, we do not know any of this with certainty. However, it has been suggested that Neanderthals created musical instruments, specifically a flute-like structure that was also a common instrument among early *Homo sapiens*. In fact, this type of instrument was almost universal in human expression and cultural evolution (e.g., Chinese flutes). Some of these surviving instruments, as previously indicated, date from around 40,000 years ago.[63]

A considerable amount of data suggests that cultural traditions were linked to music in southwest Germany.[64] The findings consistently date back 40,000 years to a rich recorded cultural history weighted toward music. Music seems to have pervaded their daily lives, punctuated by important symbol-laden rituals. Other studies have also outlined anatomical differences in neonatal brain development between *Homo sapiens* and Neanderthals, which may reflect this music/language differentiation.[65]

Music is always centered on communication, imaginative or actual, and it is deeply rooted in nature. The core of music, just like the core of language, is calling out to others. Music and language are not the same and, if Steven Mithen is correct, we may have taken the path of language while Neanderthals took the path of music. Both species, however, are bound to sounds, to an evolving auditory system in which acoustics and aesthetics culminated into song and music.

The archeological record supports the idea that humans had instruments when they arrived in Europe. What made this possible was an anatomy of walking upright, an enlarged cerebral mass underlying diverse functions, and particular social and instrumental capabilities. A richly endowed capacity spread music across diverse signaling systems. A natural predilection for music, for beat, motor skills, and intelligence, were co-opted in music making—preadaptive systems expanded in use, as is known to occur in other contexts.[66]

Language changes everything; linguistic competence, if Mithen is correct, is a major divide. We kept song, and we had musical competence. Our communicative capabilities expanded through both music and language.

Human language is an adaptive specialization in which anatomical features are designed for speech.[67] Reading and writing are much newer skills, perhaps five thousand years old. One thing that makes reading possible is gaining cephalic access to a phonological system, an aspect of the auditory system.[68] Writing-related pictures were probably an early expression, followed perhaps by expressing specific words in more abstract notational systems.[69]

Reading combines the visual and auditory systems; cephalic access to diverse adaptive systems sets the stage for reading, including the reading of music. In the evolution of reading and writing, syllables are the most accessible phoneme. Devolution of function is "phoneme inaccessibility."[70] Recapitulation of simpler systems to more complex reading and writing systems is one form of education, one form of cephalic accessibility in pedagogy. One reason we survived and Neanderthals did not may be the expansion in attention and in working memory, which makes reading and writing, as well as speech and musical lyrics, possible.[71]

Conclusion

Music emerged as part of communicative capability, a universal feature long noted and discussed.[72] Indeed, Rousseau goes so far as to suggest "that the first language of the human race was song and many good musical people have hence imagined that man may well have learned that song from the birds."[73]

Like language, the roots of music may be in the inherent shared features of our social brain, allowing us to communicate with others. Since its development, music has filled many other important roles for humans. It is a fundamental part of our evolution; we probably sang before we spoke in syntactically guided sentences.[74] Song is represented across animal worlds; birds and whales produce sounds, though not always melodic to our ears, still rich in semantically communicative functions. Not surprisingly, song is tied to a vast array of semiotics that pervade nature: calling attention to oneself, expanding oneself, selling oneself, deceiving others, reaching out to others, and calling on others.[75] These creative capabilities so inherent in music are a unique human trait.

Ian Cross, a professor at the Faculty of Music at Cambridge, has noted that facilitating the transmission of information across shared social intentional space is the pervasive social milieu; evolutionary factors are critical in understanding musical sensibility, specifying diverse social contexts in relationships.[76] We use music because it expands our communicative social contact with one another. We also enjoy music even without obvious instrumental features. Music, like other features about us, became a worthy end for its own sake.

Music is about communication. Our evolutionary ascent is the scaling of communicative competence, tracing constants of musical sensibilities to common points of origins of humanity and expansion of musical expression from this common source in prehistoric times.[77]

But musical expression is about much more than that. Musical sensibility pervades our social space and our origins in synchrony with our interactions with others that are built on core biological propensities.[78] I will return to this in subsequent chapters about music and language, in conjunction with the pleasure we derive from self-expression.

CHAPTER 2

Bird Brains, Social Contact, and Song

Song and music have their origins in biology and adaptation. Environmentally opportunistic events led to long-term viability; function predominated amid form.

Different calls serve diverse roles in varying species, but they are almost always rooted in social function.[1] The ambiance of birdsong is rich in meaning and behavioral expression; diverse animals come prepared to discern prelinguistic meaning. The range of cephalic capabilities is reflected in the richness and flexibility of semiotic expression. Evolution favored both specific local adaptations and broader, endless semiotic expression in our species, something Peirce adumbrated but did not fully engage. Semiotic capabilities, or interpreting signs, are understood in the broad array in which animals like us are endlessly interrelating with their surroundings, interpreting events and relying on cognitive capabilities.

Like human language, animal acoustic calls are semiotic, rich in information about territory, social groups, alliances, predation, danger, and resources. For instance, chimpanzees adjust their hooting sounds to the social milieu in which they find themselves.[2]

Communicative capabilities are widespread, whether in song[3] or through other forms of intimate social contact.[4] One mechanism for this is the regulation of information molecules in the brain. Two key molecules are vasopressin and oxytocin. They are not specific to music, for they underlie a variety of behaviors. Indeed, no one feature of the nervous system is particular to music. A confluence of adaptations converged to make music sensibility and expression a core and very special feature of humans—how fortunate for all of us.

I begin this chapter with a short exegesis on information molecules. Then, I connect these processes to song in frogs, crickets, and birds before coming back to neurogenesis and the working of information molecules in the human brain, focusing on some core biology underlying animal song and social contact. As we will see, steroid hormones facilitate neuropeptide expression in many species, which underlies song tied to the regulation of the internal mileu, territorial expression, reproduction, as well as a much wider range of social behaviors.[5]

Evolution and Information Molecules

Molecules such as oxytocin and vasopressin/vasotocin have a long evolutionary history. Peptides, steroids, and other transmitters can even be found in plants and insects.[6] At night, for instance, plants secrete melatonin, an information molecule that is linked to circadian rhythms in animals and acts as part of the transduction mechanism for adjusting to daylight and night time.[7] In a diverse array of animals, melatonin is linked to a number of physiological and behavioral adaptations.[8]

Plants and fungi synthesize steroids that regulate development and reproduction, respectively. Plant steroid hormones signal via membrane receptor kinases. The nuclear hormone receptors that underlie many of the actions of steroid hormones in mammals and other vertebrates, however, are absent from plants, and are estimated to have arisen more than 1 billion years ago, after the divergence of the metazoans and fungi (table 2.1).

One fundamental and unique form of nutritional and social attachment in mammals is lactation. The mammary glands, a cardinal feature of mammals, date back to Triassic and Jurassic insectivores. In fact, some birds spray their eggs with water to help nurture their development, an adaptation they retained that, on the mammalian line, may have been an evolutionary precursor toward mammary gland development.[9] Hormones such as angiotensin, vasopressin, oxytocin, and prolactin are implicated in the

Table 2.1

Examples of 'Vertebrate' Neuropeptides Found in Diverse Phyla

Protozoa	Bacteria	Fungi	Plants	Invertebrates
ACTH	ACTH		TRH	ACTH/MSH
β-Endorphin	β-Endorphin		β-Endorphin	β-Endorphin
Insulin	Insulin	Insulin		Insulin
			Melatonin	Melatonin
Vasotocin				Vasotocin

Source: Strand 1999.

maintenance of fluid balance, including the expansion of regulation that perhaps figures in the formation of new glands.[10]

In mammals, these information molecules became oxytocin and vasopressin; many of their functions were originally restricted to fluid balance, but they diversified.[11] The ancestral molecule is depicted in table 2.2.[12] Many of these information molecules are quite ancient, dating back hundreds of millions of years, although their function has changed over time.

Evolution produced the diverse functional roles, polymorphisms, and expansion in use of information molecules across diverse end organ systems. The vasopressin family is a good example of a truly antique information molecule.[13]

In molecules such as oxytocin, a kind of speciation (retaining or not retaining a common function and expanding functions) by segregation can occur. Changes in structure diversify function and affect where information molecules are expressed. Oxytocin, for example, is now expressed within auditory regions of the brain at all levels of the neural axis, and could well be involved in animals that "see by hearing" in sensory processes utilizing vital ultrasonic vocalization coordination.[14] Evolution selects the diverse sensory systems that are linked to the organization of action.

Oxytocin and vasopressin are both expressed in the auditory nuclei in several species of bats and could well be part of the

TABLE 2.2

Various Peptide Structures Across Diverse Species

Name	Peptide structure	Source
Vasopressin	CYFQNCPRGamide	Mammals
Lys-Vasopressin	CYFQNCPKGamide	Pig, some marsupials
Phenypressin	CFFQNCPRGamide	Some marsupials
Insect oxytodn/ vasopressin-like peptide (inotocin)	CLITNCPRGamide** CLITNCPRGamide CLITNCPRGamide*	*Locusta migratoria* *Tribolium castaneum* *Nasonia vitripennis*
Crustacean oxytodn/vaso press in-like peptide	CFITNCPPGamide*	*Daphnia pulex*
Vasotocin	CYIQNCPRGamide	Nonmammalian vertebrates
Arg-conopressin	CIIRNCPRGamide	*Conus geographicus*
Lys-conopressin	CFIRNCPKGamide	Leech, various mollusks
Oxytocin	CYIQNCPLGamide	Mammals
Isotocin	CYISNCPiGamide	Fish
Annetocin	CFVRNCPTG amide	Annelids
Cephalotocin	CYFRNCPIGamide	*Octopus vulgaris*
Octopressin	CFWTSCPIGamide	*Octopus vulgaris*

Source: Strand 1999.

systems for detecting auditory signals vital for approach (e.g., mother-infant signals) or avoidance in diverse species.[15]

A number of peptides include an exon duplication, loss, and exertion. Gene duplication and variation in the production of these information molecules have allowed them to play a wide variety of functions, for instance in food ingestion and metabolism.[16]

These molecules are at the heart of the arsenal that underlies the physiology of change and adaptation, as well as social attachment in social vertebrates, with both females and males participating in parental behaviors.[17]

Changes in the internal milieu over the lifespan (e.g., hormonal secretion) have long been noted. Human prenatal and postnatal development are long and varied in our species, tied as they are to the necessity of the acquisition of a huge body of knowledge during the early part of life. There is profound change in brain morphology and development during the protracted neonatal period by varying degrees in most mammals.[18] Hormones such as oxytocin are vital for parturition, the birth process, lactation, social attachment, and throughout the gestational and developmental periods.

Oxytocin is produced in the pituitary gland, the brain, and the placenta; the fetus is awash in information molecules as it floats in the amniotic fluid within the safety of its mother.[19] Learning about safety and comfort in the new and harsh external world begins in the immediate postnatal period as the neonate wails and is quickly allowed to suckle. To promote social contact, music is omnipresent in the form of maternal cooing and crooning, and perhaps reflects our capacity to learn music quite easily during critical periods in development. When one is trying to understand the behavioral response, the brain must be considered, as well as how oxytocin is regulated in the brain.

Steroids such as the gonadal hormones, or the adrenal hormones, affect the degree of neuropeptide expression (e.g., vasopressin, oxytocin) that underlies behavioral approach and avoidance systems. The genes that regulate the production of the peptide oxytocin, in combination with external circumstance, contribute to developmental outcomes.[20] They promote social attachment or approach behaviors. Regions of the brain, such as the medial region of the amygdala, underlie diverse forms of social behaviors mediated by oxytocin expression.[21] One should note that multiple factors other than oxytocin alone contribute to the diverse social behaviors of approach that we note in our everyday lives.[22]

Frogs and Crickets

The male South African clawed frog (*Xenopus laevis*) produces an elaborate song that is tied to reproduction.[23] The song facilitates communicative signaling in problem solving. Though the female frog also emits sound, it is in the form of what are called "ticks," as opposed to full song. Like birdsong, the frog song shows syntax, or a set of rules. The singing of clawed frogs is a hormonally controlled behavior and is orchestrated by androgen sites that researchers have now mapped.

The juvenile muscles that will ultimately produce frog song in adults are particularly sensitive to androgens, which induce cell proliferation in the larynx, an event that later plays a role in the expression of acoustic signals. The result of androgenic elevation is that the male larynx is three times heavier, with eight times more muscle fibers than the female larynx. Although males and females start out with similarly sized larynges, androgens in males during early development act to prevent the fiber loss that naturally occurs in females.

Investigators have observed changes in the neural circuits that control the frogs' song—notably that the cranial motor nerves experience the same hormonal regulation as the larynx during development. Using a combination of methods, investigators have uncovered the sites in the brain that underlie the singing behavior of the frog. These regions include a vocal-motor pattern generator in the tegmentum, the sensory region of the thalamus, the preoptic region, and the striatum.[24]

Investigators have studied the development of the androgen receptors that are essential for frog song, as well as gene expression and its regulation by both experience and steroid mediated events that make this possible.[25] Importantly, the steroid induction of neuropeptide regulation is a common theme. Vasotocin, an analog of vasopressin, has been linked to diverse behaviors in frogs, including mating-related behaviors and mating calls.[26]

Forebrain regions underlie mating calls in crickets as well as frogs. Infusions of vasotocin into the brain elicit song in the beginning of the call period in both crickets and frogs. Regions of the brain, including the medial region of the amygdala and nucleus accumbens, have vasotocin staining cells that are linked to these acoustic signals, and hence to song (figure 2.1),[27] as well as more generally to socially related behaviors.[28]

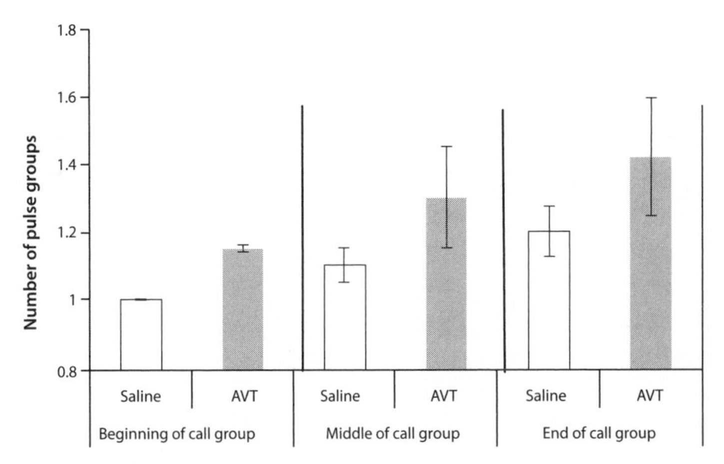

Figure 2.1 Facilitation of auditory signals following central arginine vasotocin (AVT) infusions in the cricket at the beginning, middle, and end of the call group.

Source: Marler et al. 1999.

Not surprisingly, neuropeptides such as vasotocin, vasopressin, and oxytocin in frogs or crickets are implicated in diverse forms of social behaviors, one of which is vocal communication.[29] It is linked to the quality of the environment in which the song takes place. The neuronal ensemble is quite specific and the motor/cognitive expressions that underlie the acoustic calling are transparent.[30] It appears that neuropeptides such as vasotocin impact vocal expression.[31]

Singing Birds

Birdsong is rich in melodic expression and diverse in form. Many birds (e.g., thrushes) produce multiple types of songs. Songbirds may also express different song dialects, depending upon their learning experience, context, and social terrain (see figure 2.2). Sparrows, for instance, reflect different song expression and syntax, depending on the social contexts in which they find themselves.

This diversity of birdsong reflects a great diversity of function. Hummingbirds, nightingales, and starlings sing for many different

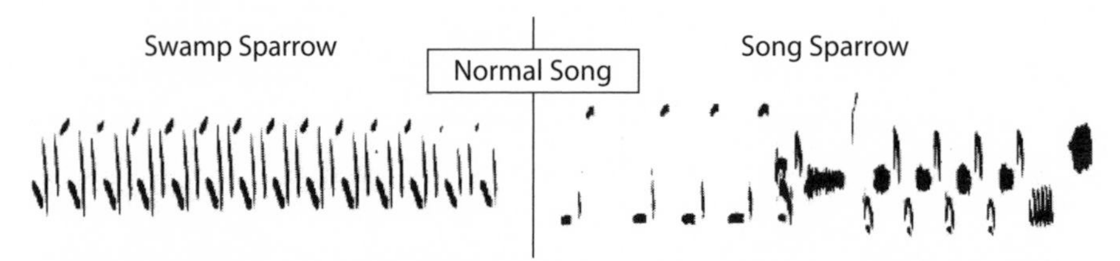

Figure 2.2 Sound spectrograms representing the normal song of two bird species. *Source:* Watson and Breedlove: The Minds Machine: Foundations of Brain and Behavior (933–36). Companion Website: Chapter 15.1, Figure 2 (After Marler and Sherman, 1983, 1985). © Sinauer Associates, Inc.

reasons (mating, predators, territory, among other reasons).[32] Exposure to song in critical periods of a bird's development is also a crucial factor in the range and expression of its song (for some birds). Song also appears to have some elements of tradition and geographical specificity; Swamp Sparrows, for instance, sing slightly different songs in varying regions of New York State.

Peter Marler, a biologist specializing in animal sounds, had a special knack for imitating songbirds, and indeed could reproduce a wide variety of the animal sounds that he studied. Having had the opportunity many years ago to take a class with Dr. Marler, it always struck me how beautiful birdsong is. These sounds are embedded in syntactical forms and embodied in brains that are quite different from ours. The songs are what C. S. Peirce called "semiotic," or meaningful, something Marler noted at the start of his career.[33]

The study of birdsong, and more generally bird watching, has a long history as a scientific hobby. In the early part of the twentieth century, Margaret Morse Nice, an ornithologist from the northeastern United States, began watching birds as a child.[34] She developed into a scientist whose research on the Northern Bobwhite and the Song Sparrow produced landmark studies, focusing on the life history of birds rather than on just descriptive accounts of their physical biology and geographical distribution. Nice was a keen student of birdsong. Birds engage in ceremonial activities, from territorial defense to food distribution, which have been elegantly cataloged and described in paintings, pictures, and clear descriptions, as well as in emerging theories regarding function.

Pinpointing function can be a scientific endeavor using scientific measures to test theories. This is not simply description, but description remains a rich and an important part of the scientific process.[35] Indeed, the description of birds is an ancient predilection. Their behavior is sometimes simple and stereotypic, and therefore easy to capture.[36] Their elegant displays of song are quite palpable and striking, which is why we have been drawn to study them. It is the aesthetic features that figure importantly into our own cultural evolution, of which music is such an important aspect.[37]

Accompanying their song, birds engage in elaborate stereotypic movements of great diversity and breadth. The orientation of these movements represents communication, as do their songs. The themes conveyed have meaning and are embedded in the generative processes of adaptive cephalic systems. Understanding song puts us at the heart of our evolution and of nature. Our natural inclination toward music is phylogenetically ancient.

We are attracted to birdsong in part due to self-involvement. Birds and humans both sing, unlike many other species.[38] In both cases music is an extension of our communicative abilities, although whether this is really "bird-song" distinct from "bird-speak" is hard to know. We are inherently social, and song is an expression of this as a specific adaptation. Music and singing are therefore liberated from a narrow adaptive feature.

While birdsong does not approach the complexity of Beethoven's works, the range and expression of birdsong is quite diverse and interesting. Predominant features include courtship, combined song and movement displays, frustration, aggression, the search for food, avoidance of predators, and reproduction. Like birds, perhaps our own joy in singing emerged as something tied to motivational systems. Then singing became something to enjoy for itself, perhaps for birds as well.[39] The means to attract and call attention, to ward off attack, to accommodate in song, now becomes something that has its own life.

Charles Hartshorne, a twentieth-century philosopher, cataloged a great number of songbirds in his book *Born to Sing*.[40] He speculated that many birds sing because they enjoy it, and that their enjoyment is bound to the aesthetics of music, as well as to its sheer pleasure. Hartshorne cited the American naturalist Wallace

Craig, whose work on appetitive and consummatory behaviors influenced a great number of thinkers, including Tinbergen and his work on instinct. Indeed, Tinbergen depicts a range of appetitive and consummatory behaviors for which we are born prepared; the appetitive is the search mode (e.g., food, water, sex, safety, warmth, social contact, and so on) and the consummatory phase is the satisfaction mode.[41] This perspective is close to the disruption model of expectations that underlies inquiry, musical and otherwise.[42]

The point for Hartshorne is that pleasure and utility are bound in the song of birds. He goes on to cite a number of musicians commenting on birdsong (e.g., Antonin Dvorak). Hartshorne, much like Marler and his colleagues, sees birds as the wondrous musicians of nature. Most humans come prepared to be naturalists with diverse categories within our cognitive structure for detecting, noticing, watching, enjoying, and for being terrified by nature.[43]

Charles Hartshorne was a philosopher by trade and merely an amateur documenter of birdsong. He is a good example of the assertion that we have a predilection to be naturalists, a trait that can express itself and evolve in suitable cultural contexts, or go in an opposite direction—the devolution of function. We come prepared to mark the contours of nature, to distinguish plants and trees. Recognizing features of nature is a biological predilection, and so is singing for birds, and perhaps for us.[44]

Birdsong is a thing of beauty. We also have a natural predilection to discern kinds of events in nature that are captivating or attractive. Biology and aesthetics are not on opposite sides; birdsong perhaps stands as a good example of this, as it is both biologically driven and aesthetically pleasing. One should not overinterpret song as a completely shared function between birds and humans, however. With the many diverse functions of music in our lives, compared to songbirds, music plays many roles for humans that have nothing to do with territory, sex, or explicit biological functions.

A favorite model system for illustrating hormonal effects on brain and behavior is singing in various bird species. Not all birds sing, but many species do, with the Zebra Finch as a favorite research subject.[45] It is quite a beautiful species; the syntax and semantics of its song compose a well-studied piece of ethology.[46] Typically, the male birds of this species sing melodic songs in order to attract females (though there are variations in the scenario, and song varies with context and experience).[47]

Birdsong depends on two events: the activation of testosterone with its conversion to estrogen in the brain, and the perception of song during critical stages of development. Testosterone also potentiates the recognition of a conspecific's song.[48] Although birds deprived of testosterone during critical stages of development may be able to sing, their song will be muted. Similarly, birds deprived of the ability to hear songs during this time by occlusion of the auditory canal will also sing only a muted song as adults.[49]

Song perception is a highly specialized ability tied to communicative and territorial competence. Like human language, song production is lateralized in the brain; the left side is dominant in these behaviors. Song production, like language production, appears to follow syntactic rules and is expressed in both male and female songbirds.[50] Song in birds resembles speech in some ways, coupled as it is to syntax and local dialects. Social interaction is a common correlate of both birdsong and speech.

Songbirds such as Zebra Finches sing during springtime, when the high-vocal-center (HVC) nucleus enlarges with elevating testosterone concentrations. Testosterone implanted surgically in the HVC nucleus facilitates song expression, whereas damage to this area abolishes or impairs song production in males.[51] Song is a cognitive motor skill depending upon context and circumstance;[52] auditory experience is a critical part of song.[53]

Song production induced by testosterone depends on protein synthesis in addition to social context. Immunohistochemical studies have indicated that a number of neuropeptides are synthesized in the circuitry underlying song production.[54] Testosterone, and the metabolic conversion process that transforms it to estradiol, induces synaptogenesis for song production in which neurogenesis and neurotrophic factors are importantly linked to season and song. It is the conversion to estrogen that is critical.

Many examples of sexual differentiation in birds (including singing) are initiated by estrogen.[55] Importantly, the brain itself is a source of the estrogen synthesis and aromatization activity that regulates song.[56] Aromatization activity is particularly extensive in the caudal neostriatum.

Estrogen implants into the high vocal center will masculinize the Zebra Finch's song repertoire. In fact, estrogen, testosterone, or converting enzymes can induce masculinization.

It is important to remember that neither estrogen nor testosterone alone controls birdsong; more important is how the two hormones are linked.[57] For example, Mark Gurney and Masakazu Konishi showed that a female Zebra Finch treated with estrogen during critical periods in development also needed treatment with testosterone for song expression when she was mature.[58]

Neurogenesis

Neurogenesis is the cellular process of procuring new neurons from progenitor cells.[59] This phenomenon was originally demonstrated in relation to birdsong and appears to be important for this essential cognitive adaptation. A great deal of evidence has demonstrated that enriched environments in both neonates and older animals engender increased neuronal sprouting. This in turn can enhance problem solving abilities.[60] Even simply enriched environments, such as those with more objects to interact with, engender greater dendritic expression in multiple regions of the brain, including the neocortex, striatum, and hippocampus (see figure 2.3 for brain regions linked to birdsong).

The degree of plasticity in cortical systems is affected by an enriched environment. Learning diverse tasks facilitates synaptogenesis in regions of the brain that include the cerebellar cortex. The formation of synapses in regions of the brain such as the hippocampus can result in long-term potentiation of this region, an area essential for learning and memory formation.[61]

Neurogenesis is something that, for some time, has been known to occur in the young. Evidence has also suggested its occurrence in the adult brains of animals, though neurogenesis slows with age. Neurogenesis is expressed in several neural sites and is tied to diverse forms of learning from others. Steroid hormones (e.g.,

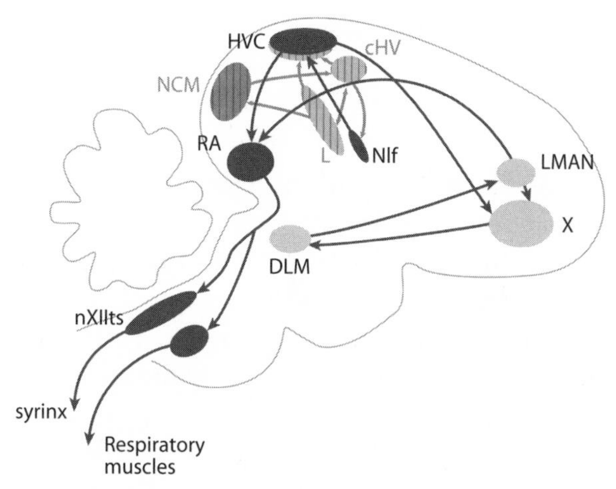

Figure 2.3 Neural sites that traditionally underlie birdsong, with associated auditory regions. The motor pathway, which is essential for normal song production, is outlined in black. In addition, the light grey pathway illustrates the anterior forebrain circuit, which is involved in song learning and plasticity. The pathway shown as a ruled oval interconnects telecephalic auditory areas, which project directly and indirectly to the song pathways. *Note*: Song system abbreviations: HVC—higher vocal center in the neostriatum (also referred to as the hyperstriatum ventrale, pars caudale); cHV—caudal hyperstriatum ventrale; NIf—nucleus interface; LMAN— lateral portion of the magnocellular nucleus of the anterior neostriatum; X—area X; DLM—medial portion of the dorsolateral nucleus of the thalamus; nXIIts—tracheosyringeal part of the hypoglossal nucleus; RA—robust nucleus of the archistriatum; NCM—caudal portion of the neostriatum; L—auditory organ in the neostriatum.

Source: Marler and Doupe 2000.

cortisol and estrogen) mediate the effects of events on neuronal expansion and reduction; they prime the brain by facilitating the expression of neuronal gene products. Neurogenesis, particularly in the hippocampus, underlies diverse forms of learning.[62] Activity can provoke neurogenesis, all of which could underlie song, dance, and musical sensibility.

Thus, neurogenesis is linked to adult learning and hippocampal function.[63] Moreover, enriched environments promote neurogenesis in the dentate gyrus.[64] One kind of enriched environment in humans is the exposure to diverse forms of music and, by implication, this could lead to forms of neurogenesis.

Song and Social Communication

Organizational and structural events related to hormonal systems are tied to sensitive periods that facilitate the organization, and the later conditions for the activation of diverse behaviors essential for successful adaptive responses. They occur both pre- and postnatally. Lifelong change, maintenance, and evolution are features of the brain, supported by diverse steroids for long periods (e.g., estrogen promoting tissue, corticosterone, deteriorating tissue, and by the induction of neuropeptides like growth factors, oxytocin, or vasopressin).

A conversion process that transforms testosterone into an estradiol by an enzymatic process is truly remarkable as a piece of natural history.[65] In other words, testosterone is converted to estrogenic compounds that have long-term structural impacts on the brain and on behavior; birdsong is but one example. Song is linked to aromatization, the conversion of testosterone to estradiol, in many species of songbirds.[66]

Song plays diverse roles in birds, including social calls, calls about predators, food resource indications, and mating calls. Song is syntactically structured and formed during critical periods in which acoustic signals are vital, in addition to steroid conversion for the expression of adaptive adult song.[67] However, song is almost by definition social; it is a call to others, maintaining a rich semiotic component.[68]

The meaning of song has been a cottage industry for the professional semiotician over the last hundred years or so, as well as a hobby for the amateur. The naturalist, merging with the experimentalist, looks at the interaction between the theory and the test, varying the song at critical stages in the development of the young bird that is exposed to it to determine whether song is a function of critical exposure. For some species of bird and types of song, it is, and expectations about the songs of conspecifics permeate dual and coordinated singing in diverse singing birds.[69]

This critical exposure provides a framework of expectations that configure the structure of song to the generator and to the listener. On the semiotic side, the listener is a bit like the pragmatist inquirer. Expectations predominate until an expectation is thwarted. Attention is promulgated forward to a set of orienting responses in which to discern a disruption of cognitive motor habits. After all, this sort of biological adaptation is cognition in action.

Thus, gonadal steroids are vitally involved in song production in birds, starting with the presence in the yolk of maternal testosterone in developing birds.[70] This may increase vasopressin in diverse regions of the brain that underlie song, as well as increasing sexual and territorial aggression.[71] Again, the behavioral effects of the steroid can be rather rapid via membrane-related effects.

While testosterone induction of vasotocin facilitates song production in several bird species, estrogen also has a profound effect on the brain, with organizational implications that promote changes in structure and function essential for diverse social behaviors. Gonadal steroids then trigger vasotocin or vasopressin, which have different effects depending upon the species and context.

Vasotocin, Vasopressin, Song, and Communication

As was already noted, vasopressin, like oxytocin, is phylogenetically ancient and linked to diverse functions, including osmotic and fluid regulation.[72] Vasopressin is richly expressed in the brains of all vertebrates studied to date.[73] One important example of its activity is its link to gonadal steroid hormones during critical periods of development via expression in the brain.[74] Vasopressin is traditionally associated with pituitary release in the control of water volume at the level of the kidney when water balance is compromised; when water balance is depleted, pituitary vasopressin acts within the kidney to facilitate water conservation.[75] But kidney regulation is only one function of vasopressin. As a neuropeptide, it takes many diverse roles.

One of its functions is in the maintenance of territorial defense, inducing defense aggression related to external circumstance and sustained by gonadal activation. It is also active in the induction of behavioral competence, for example, in the territorial aggression that is required to maintain control over physical space.[76] In addition to critical periods setting down conditions of vasopressin in diverse regions of the brain that underlie social behaviors, the expression of vasopressin is under steroid regulation.[77]

"Social calling behaviors" are, of course, not unique to birds. There are changes in the brain in other species reflecting similar hormonal changes.[78] The steroid secretion sets the conditions for facilitating structural changes that are then activated across the life cycle. Some of these events occur in utero, from temperature or warmth regulation, to events that facilitate song production and perception.[79]

One role for aromatization is inducing structural changes in peptides and neurotransmitter levels in the brain.[80] A result is the altering of diverse information molecules such as peptide hormones that underlie behavioral adaptation; the alteration involves not only the structure of particular nuclei, but also the production of information molecules.[81] Vasopressin is but one example and is perhaps linked to natural song signals about territory, about conspecifics.[82]

Vasopressin and Social Behavior

Song is a social behavior, as are other expressions of musical sensibility. Affilitative behaviors are fundamental to a variety of species, and in some species with long-lasting affiliations (e.g., monogamy) they are linked to differences in neuropeptide expression (e.g., oxytocin, and vasopressin).[83] Slight changes in the vasopressin gene may underlie these differences in social behavior (figure 2.4).[84]

Steroid hormones such as estrogen or testosterone are essential for sustaining neuropeptide expression and receptors. They underlie the behavioral regulation for approach/avoidance behaviors. Vasopressin 1a receptors underlie maternal defensive aggression, mediated by estrogen levels.[85] Vasopressin also underlies social communicative functions that include human social contact and affiliated behaviors.[86]

There is now evidence to suggest that changes in vasotocin expression, facilitated by testosterone in the frog or songbird, may

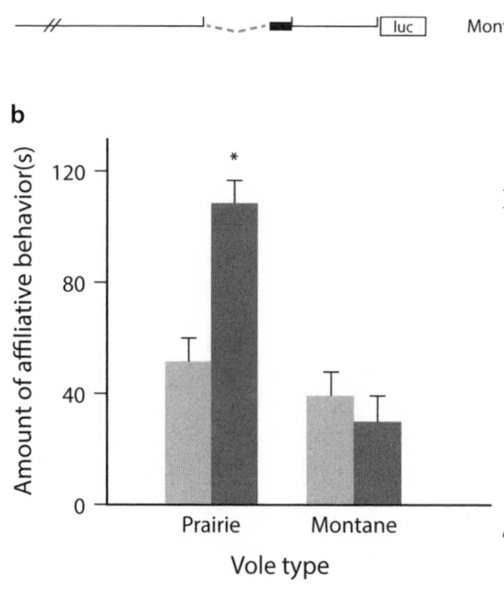

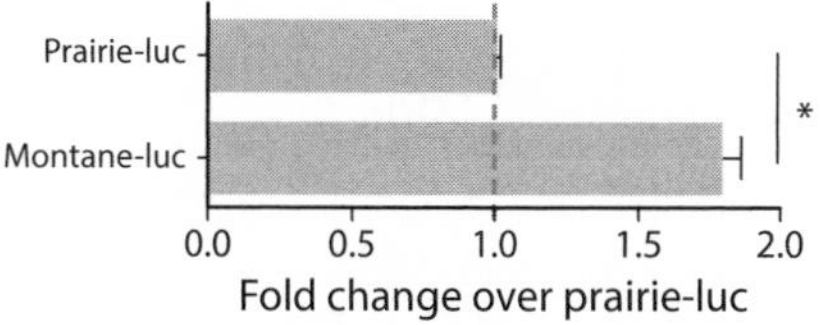

Figure 2.4 (a) Species differences in structure at the microsatellite region: functional microsatellite polymorphism associated with divergent social structure in vole species. (b) Increased affiliative response in voles following administration of AVP compared to saline.

Source: Hammock and Young 2004.

underlie song.[87] For instance, at some time periods during the year, vasotocin is known to enhance the singing behaviors of canaries in steroid-treated animals.[88]

Moreover, vasotocin infusions into the third ventricle of the brain of White-crowned Sparrows facilitated vocalization (in addition to non-song behaviors) (figure 2.5); none of the animals that were tested sang without this treatment. The effect was particularly strong in estrogen-primed females.[89]

The distribution of binding sites for vasotocin in several songbird species occurs across various regions of the brain.[90] Vasotocin is also tied to vocalization in diverse species.[91] One example is the effect of estrogen on oxytocin gene expression. In many species, estrogen is known to facilitate oxytocin secretion and different forms of maternal behavior.[92] It is the induction of oxytocin gene expression that underlies this complex behavior, especially in limbic regions of the brain including the hypothalamus, amygdala, and bed nucleus of the stria terminalis.[93]

Of course the social milieu also affects end organ systems such as the brain, including the regulation of vasotocin or oxytocin.

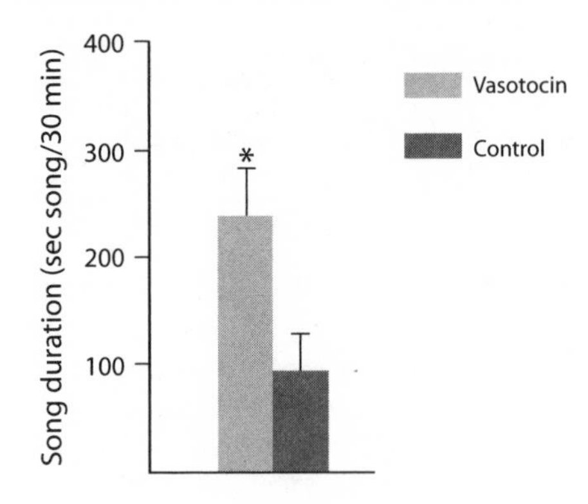

Figure 2.5 Facilitation of song in males canaries by vasotocin.
Source: Voorhuis et al. 1991.

That is, the social milieu helps determine the level of gene expression in the brain. It is vasotocin or oxytocin that is importantly involved in sustaining the behaviors associated with social communication and social attachment behaviors.[94]

Social Contact, Vasopressin, Oxytocin, and Amygdala Function

Vasopressin and oxytocin are ancient hormones and, as noted, they are tied to fluid regulation as well as successful social contact across many species, successful gestation by regulation of steroid hormones, and the induction of oxytocin expression in the placenta and perhaps the brain.[95] The range of behavioral events linked to the central release of these two neuropeptides is depicted in table 2.3.[96]

Oxytocin and vasopressin also have rich homological histories, as depicted in the next figure. A slight change in the receptor structure (in this case the vasopressin 1 receptor subtype) can have a profound impact on our social behaviors (see figure 2.6).[97]

Oxytocin, like vasopressin, is bound to many forms of biological/behavioral regulation important throughout the life cycle (table 2.4).[98] Social attachment has short- and long-term regulatory functions essential for one's lifetime.[99] One adaptation is social contact, which is vital for musical sensibility.

TABLE 2.3

Depiction of the Many Functions of Vasopressin and Oxytocin

Social Stimulus		Social Response
Parturition	AVP/OXT	Maternal care
Suckling/Pup vocalization	Central Release	
Intruder conspecific		Aggressive behavior
Resident conspecific		Defensive behavior
Mating partner		Sexual behavior
		Social bonding/Affiliation

Note: AVP = Vasopressin
OXT = Oxytocin
Source: Veenema and Neumann 2008.

Neuropeptides are more specific than classical neurotransmitters with regard to behavior.[100] Music expression is an ideal behavior as it involves social contact, the promotion of positive social sensibilities, social attachment, and a sense of well-being.

Social attachments are footholds by which we are connected into the larger world and anchored to others. So we require, as it were, a social network of genes geared toward social attachment, which include oxytocin and vasopressin underlying part of the social brain.[101] Thus, oxytocin with regard to auditory signaling is an element of social attachment that may provide a foothold into the world of the song of the mother to the neonate.[102] The neonate bias is toward a familiar sound, quickly shaped by the language and song that he or she is exposed to, which no doubt can have an impact on cephalic expression and core physiological responses (e.g., cortisol secretion).

Regions of the brain, such as the amygdala, long linked to reward and social or song perception, are also tied to social judgment or quick heuristic judgments.[103] Such fast judgments underlie many social decisions, and diverse forms of cognitive adaptations emanate from a prepared system in which information molecules such as oxytocin, vasopressin, or vasotocin are critical for social contact and perhaps musical expression.[104]

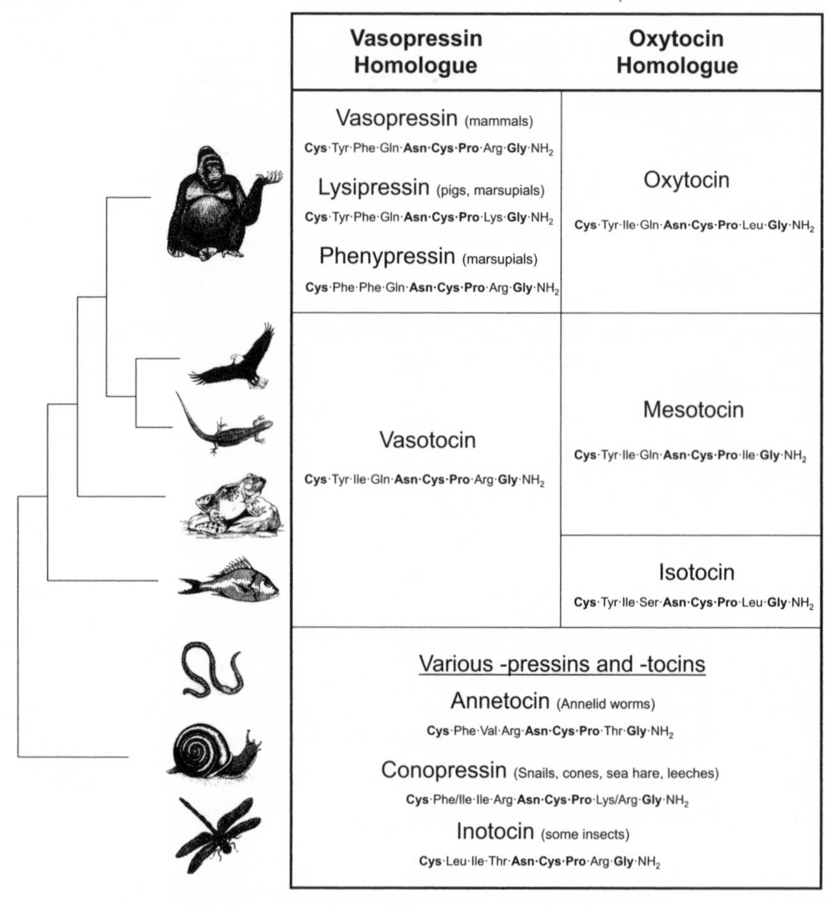

Figure 2.6 Oxytocin and vasopressin homologs found in diverse species.
Source: From *Science*, Nov. 1, 2008. Donaldson, Zoe R. and Larry J. Young, "Oxytocin, Casopressin, and the Neurogenetics of Sociality." Reprinted with permission from AAAS.

Underpinning a diverse set of cognitive adaptations are rich sensory/motor systems endowed with the ability to understand the gist of things and be linked together.[105] Some of these adaptations are tied to amygdala function. The amygdala is linked to many forms of social contact and is broadly associated with musical sensibility (see chapters 4–6).

In addition, regions of the amygdala essential for social attachment and avoidance also demonstrate significant changes in humans: for instance, enlargement of the lateral amygdala, which is

TABLE 2.4

Bio-behavioral Functions of Oxytocin

OXYTOCIN

↓

Positive social interaction

↓

Activation of sensory afferents by non-noxious stimuli

↓

OXYTOCIN

↓

Stimulation of anti-stress effects and growth

↓

Stimulation of attachment or bonding

↓

Repetitive sensory, mental, and conditioned stimuli of non-noxious type

↓

OXYTOCIN

↓↓↓

Sustained anti-stress effects and stimulation of growth, comfort, and musical expression

↓↓↓

Promotion of health

Source: Uvnäs-Moberg 1998.

closely tied to neocortical function.[106] The largest nuclear region is the basal lateral region. In one comparative study of humans and apes (e.g., chimpanzee, bonobo, gorilla, orangutan, gibbon), investigators found that the size of the lateral division of the amygdala expands significantly in *Homo sapiens* compared to the expansion in other primates (figure 2.7).[107] This region of the brain is

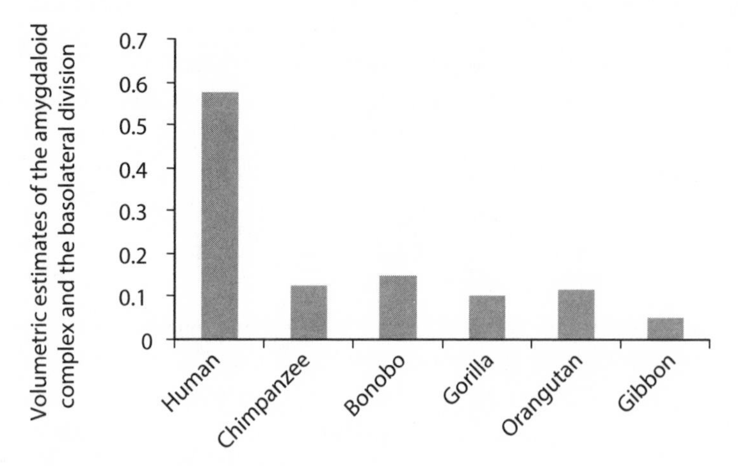

Figure 2.7 An evolutionary trajectory illustrating a volumetric increase in the lateral region of the amygdala.

Source: Barger et al. 2007.

consistently linked to expectations, social regulation, musical sensibility, and broad-based affective responses.

Moreover, the amygdala, among other brain regions, underlies the perception of others—their intentions, beliefs, and desires. For instance, facial perception and eye contact are linked to amygdala function.[108] This ability to understand the beliefs and desires of others is fundamental to our social knowledge and our ability to be anchored to the social world of others. Vasopressin administered in humans through nasal inhalation activates a number of regions in the brain that may underlie our ability to recognize the beliefs and desires of others.[109]

Since social contact in humans is achieved in part through visual representation, it is not surprising that oxytocin is linked to shared functions and social attachment.[110] Social contact, including physical closeness, eye contact, social listening, and so on, provides an important part of our well-being, and underlies all aspects of musical sensibility. Oxytocin and vasopressin are two closely linked information molecules tied to social contact, with persuasive evidence suggesting that they might be associated with our musical sensibility as well.

Human studies show that oxytocin inhalation facilitates social contact, recognition, and social memory.[111] Importantly, higher oxytocin levels are associated with increased likelihood of reciprocation in a game in which trust is quantified: the higher the levels of oxytocin, the greater the trust. Oxytocin levels were found to be higher in subjects who received a monetary transfer signaling an intention to trust, in comparison to an unintentional monetary transfer of the same amount from another player.[112]

In another study, investigators noted that intranasal administration of oxytocin induced more cooperation in an anonymous economic game by boosting interpersonal trust and impacting amygdala activation—a feature that underlies musical sensibilities.[113]

Conclusion

This chapter suggests that a core set of neuropeptides acting on diverse regions of the brain underlies communicative song—song being essentially tied to social context and social transmission of information. Experiments with birdsong demonstrate that song production in songbirds is intricate and these studies show that song production and full song expression reveal several properties of importance:[114]

(1) There are sensitive periods in which the hearing of song is important for the later expression of song production;
(2) Song production reveals a form of syntax and creative variation;
(3) Song is essential for successful reproduction in songbirds; and
(4) Song syntax systems are lateralized to the left side of the bird brain, analogous to syntax and language lateralized to the left side of the human brain.

I am not suggesting that birdsong is more like human language than human music; indeed, birdsong overlaps with both forms of human expression and at the same time is endlessly different.

Nevertheless, as we will see, many of the same neural circuits that underlie language in humans also underlie music.

Song production and song perception presuppose a rich cognitive system in the brain of these birds that serves diverse functional roles (e.g., territory, mate attraction). Darwin noted that birds may express emotions through song.[115] In other words, birdsong presupposes an innate template of possible syntax structure, over which semantic content is conveyed to conspecifics.[116] But song appears aesthetic to us; is it to birds? We do not know. It certainly can be aesthetically pleasing to our ears. We do know that communicative function, the full form of expression, incurs biologically greater increases for successful reproduction, social contact for which oxytocin and vasopressin are a common neural currency in the organization of action. Song is part of the communicative interaction between living things that sing and understand the song, using the information in functional contexts (approach, avoid, find, and so on). There is no knock-down evidence for the attribution of aesthetics to other species, but I do not want to unnecessarily distract the reader from the strong sense of the importance of aesthetics in us. I do suggest that aesthetic sensibility evolved in the context of information processing systems and adaptive responses to the world of which we are trying to make sense.

The unique phonological combination from a source syntactical template in birds, however, does suggest a possible precursor for song production in humans.[117] Understanding acoustic signals and their transduction is fundamental to auditory physiology.[118] It underlies song and syntax. Calling out to conspecifics, warning, attracting, and deceiving others in diverse semiotic contexts, are pervasive features of nature.

Thus, research on birds, though intrinsically interesting, does also have bearing on understanding human song and language. Between birds and humans are homologues of brain regions, and closely related parallels of syntax such as left hemispheric dominance, but the differences are also vast, despite the beauty of their similarities (figure 2.8). And while some similarities exist between birdsong and human expression, the differences vastly outweigh the similarities.

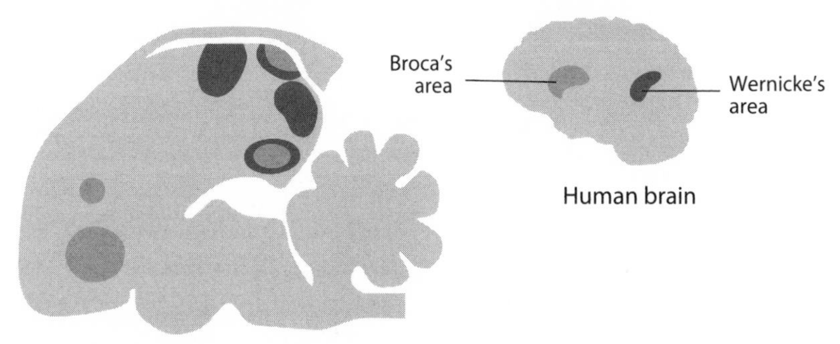

Wernicke-like regions activated when the bird hears song

Broca-like regions activated when the bird sings

B

Figure 2.8 (a) Common convergence in evolution; Broca's and Wernicke's areas, regions specialized for vocal learning, in bird and mammal brains. (b) A Zebra Finch.

Source: (a) Balter 2010; Bolhuis and Gahr 2006. (b) Wikipedia.

CHAPTER 3

Human Song

Dopamine, Syntax, and Morphology

Rhythmic beats are literally at the heart of us. The syncopated heartbeats of the mother and the infant are routinely in rhythm with each other. The rhythmic babble of the infant turns to formed speech; the syntax (structured movement into predicted sequences) underlying movement is initiated and sustained by internal generators coupled with events from the external world.[1]

A number of cephalic adaptive systems underlie the orientation to song and music, but they are for the most part not specific to song. Oliver Sacks, a clinical neurologist and musician himself, has famously described the role of music and song in the treatment, and sometimes the recovery, of patients with Parkinson's disease or those who suffer strokes.[2] Sacks's work demonstrates how profoundly our aptitude for music is tied into the way our brain works, in addition to our well-being.

Many different cognitive and motor systems fuel our ability to organize action, thought, and music. This chapter's focus is more on the motor side. As Steven Brown has suggested in his review of A. D. Patel's very impressive book on music, motor systems and motion have not been emphasized enough in the study of music.[3]

In this chapter, I look at the fundamental link between cognitive systems, movement, and the organization of the brain; what Karl Lashley, a noted psychobiologist of the twentieth century, emphasized in his essay on the "serial order of behavior" that also discusses music.[4] Musical expression is just one example of serial orders of behavior and diverse human creative expression. I begin with the ties between speech, song, and language in the evolutionary context, and then move on to considerations regarding the neurotransmitter dopamine, a vital chemical for syntax, song, movement, and the prediction of events.

Speech, Song, and Language

We do not know exactly when humans began to speak, but we do know it probably happened tens of thousands of years ago. All human language seems to have a common origin, spreading quickly. Language is coupled to a number of evolutionary paths deriving from social context.

Language is, in part, a social tool, with cognitive/motor systems forming the foundation of its use. Some linguistic theorists even speculate that language's origins lie in the need to gossip or to spread news about others' social transgressions as a means of keeping track of social alliances and of hierarchical social relationships.[5] Human lexicons have great diversity and enormous generative capability.[6]

Language is also melded to an evolving motor system. After all, Broca's region, where many speech functions appear to be linked, is a paradigmatic motor-related area. It is tied to syntax, but also action, as well as language in action.[7]

Some experts locate the origins of speech in gesture. Even the apparently simple gesture of pointing is rich in meaning, a window into the mind. This gesture is joined to speech and language in general and yet remains something separate. The information of a gesture is not the same as speech, but an adaptation that enhances our social communicative prowess and serves as an expression through the motor systems.[8] Motor control cuts across the whole of the central nervous system.[9]

Consider the pointing of the conductor bringing together the diverse instruments and people in an orchestra, to quiet or accentuate, integrate, exaggerate, or facilitate the melodic tapestry of harmonic and dysharmonic moments. Inherent in the conductor's pointing is a plethora of cognitive systems linking with others—and simply by pointing.[10] Each of the gestures made by the conductor, although ostensibly just pointing, is in fact unique. And so gestures are also distinctive and multifaceted, like speech.

Cognitive motor systems, like those behind pointing, are embedded in language and gestures. Gesture is linked to speech because it is an important prelude to syntax and modern language expression. Gesture itself may be a form of language.[11] The knowing process glides through the physical sense.[12]

Gesture permeates the social signaling systems, as well as music. Watch the bass player "making love" to the bass, the physicality of the clarinet and saxophone players, the drummer, the song and melody reflecting the pervasive presence of gesture in musical sensibility. It is the body, in Merlau-Ponty's words, "haunting" the space of interaction; gesture pervades that space. As Michael Corballis noted in *Hand to Mouth*, gesture is tethered to language, a cephalic capability that changed the landscape of human action, human meaning, and human social exchange.[13]

Gestures, the Brain, and Syntax

Some cultures are more gestural than others. I grew up with many Italian Americans and I am forever waving my arms when I talk, using diverse gestures, some not so nice, but always expressive and meaningful. The cognitive motor function is imbued across the terrain of the gesture. It provides greater diversity of expression, with more meaning in a landscape of semiotics.

Semiotics is rooted, as C. S. Peirce noted,[14] in reference and generally in meaning. This is well represented in musical semiotics replete with gesture. Gesture underlies the broad array of musical expression, learning, and performance, and it is inherently rich in social communicative contact.

Semiotics, the sense of meaning that pervades communication, Peirce thought, are never as precise as one might want; vagueness and generality are ingredients in the perceptual availability of everyday transactions. Signs that serve as icons, indices, and symbols permeate the brain's social space. These are rich bearers of information in which the knower and the known are in contact. This is certainly true in music. There is no Cartesian separation in a fabricated rationalistic space, only engaged activity, the signs rich in scaffolding toward conceptual sanity and expansion.[15]

Forging links between events is one essential feature of our cognitive efforts. Semiotics is the connective glue that helps us to understand the links between an image or symbol and meaning.[16] Semiotics occurs in an individual (e.g., learning to read music), but is also transmitted culturally and plays an important role in our ability to cooperate socially (e.g., the meaning of traffic lights). A sense of semiotics, Peirce thought, is something "virtual" in the bounds of information and human connectedness, building into a larger public space of meaning.[17] This is true whether we are talking about understanding, our sense of music, or our sense of inquiry about other species.

Movement and action, rhythm, directionality, and purpose pervade most music. At the same time, expectations are inherent in music. Even in an improvisational jam with others, there is order and pattern. Improvisation requires the use of tones that break traditions, starting new ones; but the movements are nevertheless somewhat predictable. Without this "contract" of predictability, this social action could not occur. These constants are transmitted through semiotics. Leonard B. Meyer, a "fiddle player" as he described himself, as well as a composer mentored by no less than Aaron Copland, points out this same idea.

Part of what allows us to improvise musically with others is that we are endlessly social. Our eyes are always on each other. Human gesture also involves the same dance of contact and understanding, a social milieu diverse and wide-ranging, embedded in cognitive/motor coordination. Motor skills and language are fundamentally linked. Thinking of action words (arm or leg words for instance) will activate motor regions of the neocortex; this requires central dopamine regulation, which also underlies musical expression. Gestures are ripe with symbolic meaning, so that even when we are not talking, we are communicating with one another through eye contact, shoulder shifts, and other similar motions. This is key to musical development and also to our sense of social connectedness in almost any setting.

Dopamine, Neural Circuits, and Behavior

Pointing and other actions are tightly coupled to thought, with dopamine regulating them. Indeed, central dopamine is a fundamental neurotransmitter that underlies the organization of effort. Dopamine is a catecholamine produced in the adrenal gland, as well as in core structures of the brain. Dopamine is just one neurotransmitter among others, but it seems to be particularly important for the organization of cognitive systems and for action.

Dopamine appears to be a fundamental neurotransmitter for language, and probably also computations. It is central in the organization of drives and rewards.[18] Dopamine may be crucial to the development of behavioral inhibition as well.[19] Both excess and depletion of this neurotransmitter are reflected in diverse forms of pathology (e.g., Parkinson's disease). The regulation of dopamine is fundamental for behavior. It is an ancient molecule dating back millions of years in evolutionary history and plays a critical role in the motor control of the nervous systems of all vertebrates.

Dopamine, serotonin, and norepinephrine are all amines. They are represented in specific neuronal sites in the brain stem and forebrain. From small clusters of cell bodies, a diverse array of fiber pathways connects to a large part of the brain; this is true of all amines (e.g., dopamine). For example, dopaminergic pathways from the substantia nigra innervate the putamen and caudate nucleus of the striatum; these same regions of the basal ganglia facilitate prosocial synchronization.[20]

The nigrostriatal system is involved in subcortical motor control. Dysfunction of this system is associated with certain movement disorders, such as Parkinson's disease and tardive dyskinesia. The initiation of movement and the organization of thought are important functions of the nigrostriatal system, and dysfunctions of this system are illustrated by the characteristics of Parkinson's disease.[21] In other words, there is an associated associated decrease in certain aspects of cognitive performance along with the well-known movement impairments in Parkinson's disease.[22]

The mesocorticolimbic dopamine system includes the brain systems that are important for the ingestion of food, water, drugs, social reinforcers, and other rewards. This system can be subdivided into at least two networks: the mesolimbic system and the mesocortical system. The mesolimbic system consists of dopaminergic projections from the midbrain ventral tegmental area to the limbic forebrain areas, including the nucleus accumbens, stria terminalis, lateral septal nuclei, amygdala, and limbic areas of the striatum.[23] In the mesocortical dopamine system, the dopamine cell bodies are also located in the ventral tegmental area. Subsets of dopamine projections terminate in the prefrontal cortex, anterior cingulate, and entorhinal cortex.

Given its wide involvement in brain areas and systems, it is not surprising that dopamine levels are linked to diverse motivated behaviors.[24] These links have led a number of investigators to connect dopamine to reward. However, dopamine neurons are activated under a number of conditions, including duress or excitement. The pain of performance rituals through rehearsal and the expected excitement of the musical experience in context with others, for instance, both activate dopamine.

With regard to the brain more generally, regions of the striatum are now known to underlie appetitive motivation. Dopamine expression plays an important role in the organization of appetitive behaviors. Dopamine is also involved in generating motivated behaviors, no doubt including musical behaviors (performance, perception, syntactical, and affective connections). For instance, the range of tonal expression is reduced in Parkinsonian patients.

This vital neurotransmitter is also central to the organization of movement and is in control of sensory information. Depletion of dopamine compromises behavioral competence and sensory/motor integration, which contain the basic ingredients for behavioral control.

The expression of central dopamine influences the attentional response to environmental information. It decreases or increases attention to, and the learning of, environmental events. In other words, increased salience, a key factor in musical sensibility for both the perception and the integration of environmental information, is one consequence of elevated dopamine levels.[25]

Neurotransmitters such as dopamine do many different things. For instance, depletion of central dopamine in dopaminergic regions of the brain is linked to impairment in movement, long noted in Parkinsonian patients.The same region is also linked to syntax in language.

Dopamine is also tied to the organization of thought and action. In addition, there is some evidence that dopamine expression can benefit from rhythmic-auditory stimulation, which facilitates better movement and expression in Parkinsonian patients.[26] Music, as we will see, enhances dopamine expression as well.

Dopamine is neutral with regard to function; it is only essential for the organization of behavior or the organization of effort (e.g., solving a math problem, running, persevering at practicing the clarinet, inhibiting anti-social behavior). It underlies the feeling of effort (practice, practice, practice, and yet more practice) and the rational prioritizing of our goals. Dopamine is active, I suggest, under both positive and negative conditions; for instance, either when one approaches something wanted or needed, or when avoiding something aversive, dopamine is involved.

Musical performers are but one example of this exercise of effort. A serious competitive swimmer is another paradigmatic example of perseverance through physical and mental adversity. Diverse cognitive resources are embedded in both musical performance and sustained anticipatory behaviors to reach out to the audience, the social milieu. Of course, musicians and swimmers both have to balance a sense of reward with the pain that they might be experiencing. They have to withstand short-term discomfort and set their sights on anticipatory, longer-term satisfaction.[27]

Music, no less than swimming, is action oriented, whether literally in the movement of the virtuosity of Liszt, or in the controlled building up to a crescendo and release as in "The Lark Ascending" by the twentieth-century composer Vaughan Williams.[28] Action permeates music, and dopamine underlies the action of thought and the diverse cognitive systems that orchestrate the embodied expression of music.

Dopamine, Discrepancy, and the Prediction of Reward

It has long been known that neurotransmitters in the brain play an important role in learning and reward. There is no univocal conception of reward, and I use it in a commonsense manner—something desired, something labored for, something attained. Dopamine provides an excellent example for the role of the body in the prediction of rewarding events, which may underlie aesthetic judgment.[29] Dopamine is involved in both action and thought; as such, it is a necessary chemical information molecule for us in maintaining a coherent world in which to function.[30]

The dopamine pathways in the brain underlie a number of behavioral functions that range from syntax production and probability reasoning, to the activation and learning of specific motor programs.[31] For example, Parkinsonian patients (who have depleted dopamine levels in the brain) have trouble with ordinary syntax as well as with motor control.[32] The mechanisms for generating action are not separate from the mechanisms for thought. The serial order of behavior is organized by a number of regions of the brain, including the basal ganglia, in which dopamine is a primary neurotransmitter in the organization of behavior.[33]

An interesting set of studies on dopamine neurons in the brains of macaques has suggested that one function of this neurotransmitter is the prediction of rewarding events (such as hearing music);[34] dopamine neurons tend to fire more in anticipation of rewarding events.

Dopamine neurons are linked to the learning phase in a number of paradigms and in response to novel events.[35] Dopamine neurons in the frontal cortex and striatum are active in anticipation of these events (e.g., gustatory, auditory, and visual rewards).[36] In electrophysiological studies in macaques, before a reward like a nut becomes fully predictable, dopamine neurons are activated each time the reward is given. Once the reward is predictable (i.e., always occurring without failure), dopamine is no longer activated to the same degree. However, if the predicted reward does not occur as expected, dopamine release is depressed (figure 3.1).

The unpredictable feature of a stimulus is importantly linked to dopaminergic activation, reinforcing the Rescorla/Wagner model

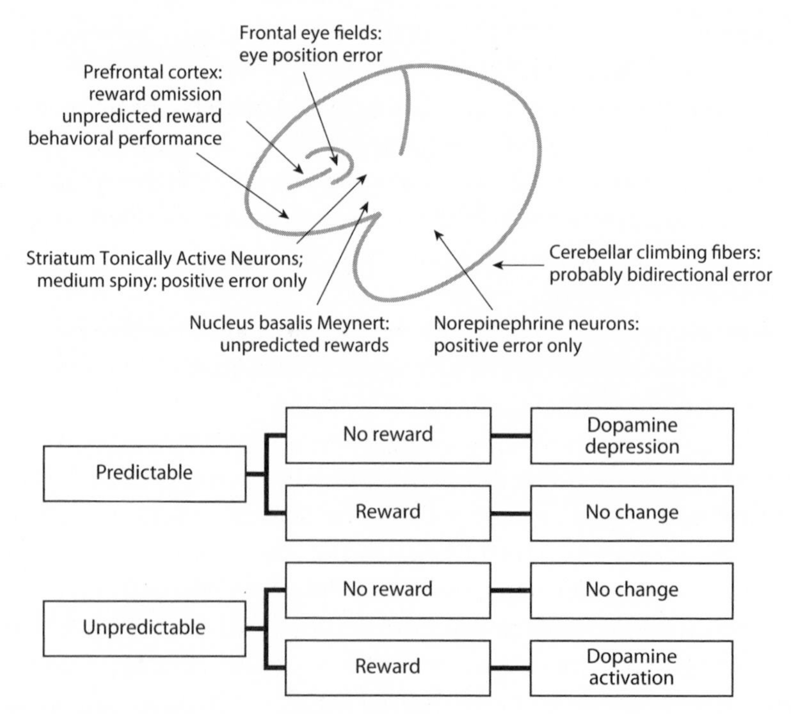

Figure 3.1 Regions of the brain linked to dopamine, and their putative activation in predictable and unpredictable events.

Source: Schultz 2002.

that predictability is a primary feature in expectancy, attention, and learning (see chapters 4, 5).[37] In further studies, representations of reward predictability or uncertainty were correlated within populations of dopaminergic neurons. The greater the uncertainty of reward occurrence, the greater the activation of the dopamine neuronal population.[38] Two sets of neuronal populations are responsive to these events: one set changes to reward probability, while the other changes to reward uncertainty.

The important point is that this is a model of expectancy and learning, and the model of dopamine release should have some application to aesthetic appreciation. What is expected and easily predicted creates less of a reaction than something that is novel or unique. This again does not mean that it is more pleasurable.

Dopamine is neutral on that score. For instance, many studies have shown that facial symmetry is aesthetically appealing. This may mean that people do not like violations of expectations. The same holds for the perception of beauty or aesthetics,[39] which creates a definitive reaction in the brain, through the dopamine neurons.[40] So how does this relate to our understanding of music? Could it be that when there is an unexpected twist in a musical score, we sense that as rewarding?

Dopamine has long been linked to issues of reward as well as the liking of and approach to certain objects.[41] The concept of reward, however, has a long and thorny history. Dopamine is not simply a neurotransmitter underlying the brain mechanisms linked to reward. It is much more complex—even when dopamine is blocked, animals can still "like" things (e.g., sucrose). Indeed, dopamine is more tightly linked to the motivational component of pleasure-related events, and can be separated from the predictive reward components, while some of the endorphins are linked to the ingestion of a reward.[42] Thus, importantly, dopamine is not only essential for the organization of drive, wanting something, but also for the salience of events (figure 3.2).[43] In other words, dopamine expression is essential for incentive motivation—the range of associations that is generated in terms of objects to approach or objects to avoid.

While dopamine may not be necessary to enjoy something, it is necessary to motivate an animal to seek something. The induction of dopamine in regions of the brain underlying the organization of action increases the propensity to perform a number of functions, including music. These regions include the basal ganglia, of which the nucleus accumbens and ventral palladium are a part, which receive dopaminergic projections from the ventral tegmental area. Interestingly, using fMRI as a measure of brain activity shows that the activation of the nucleus accumbens is a predictive factor in the purchasing of popular music in the United States (figure 3.3). In other words, the greater the activation, the more likely the music will be bought.[44]

The dopamine system does not function in isolation. Dopamine is co-localized with a number of other neurotransmitters and

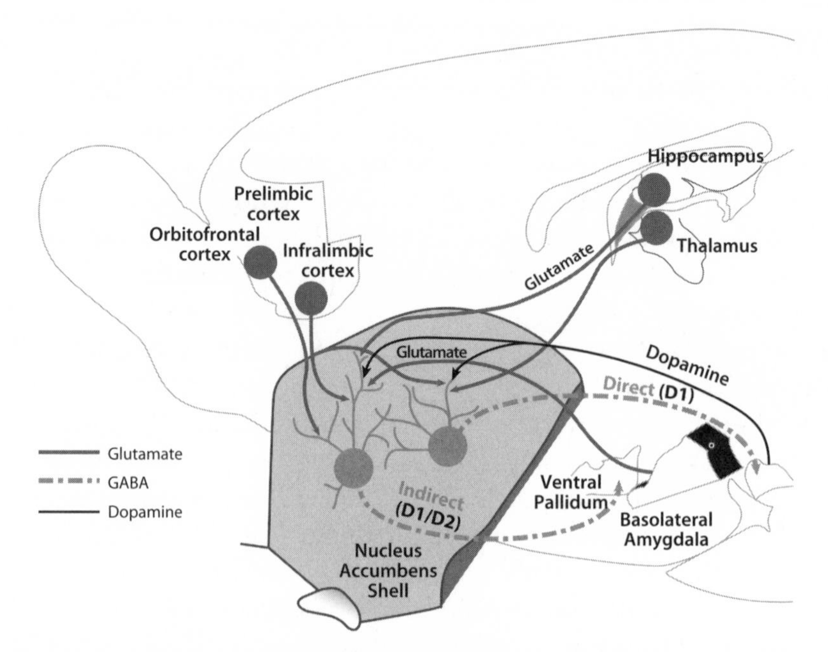

Figure 3.2 Regions of the brain that underlie incentive salience, in part through neurotransmitters such as dopamine, glutamate, and GABA.

Source: Richard and Berridge 2011; Faure, Richard, and Berridge 2010.

neuropeptides at a variety of sites in the brain.[45] Neurons in the nucleus accumbens receive input from the hippocampus as well as from the brain stem (serotonin neurons and glutamate neurons), the cortex, and the thalamus. Neurons within the nucleus accumbens contain receptors for many other neurotransmitters.[46]

Dopamine is just one neurotransmitter in the context of the organization of action. What is interesting is that this one neurotransmitter, so essential in the organization of movement, is also essential in the organization of thought—namely the prediction of events. The prediction of events is one cognitive component that underlies a wide variety of our activities, including aesthetic judgments. One mechanism underlying aesthetic judgment of music may require (and this is speculative) dopamine neurons in rule-governed expectations that underlie grammar and syntax.[47] Cognition and bodily sensibility to art (e.g., music, a painting, a ballet, a sonnet) are not opposites; they run together. Indeed, as Meyer understood, cognitive systems run through the bodily expression within music.

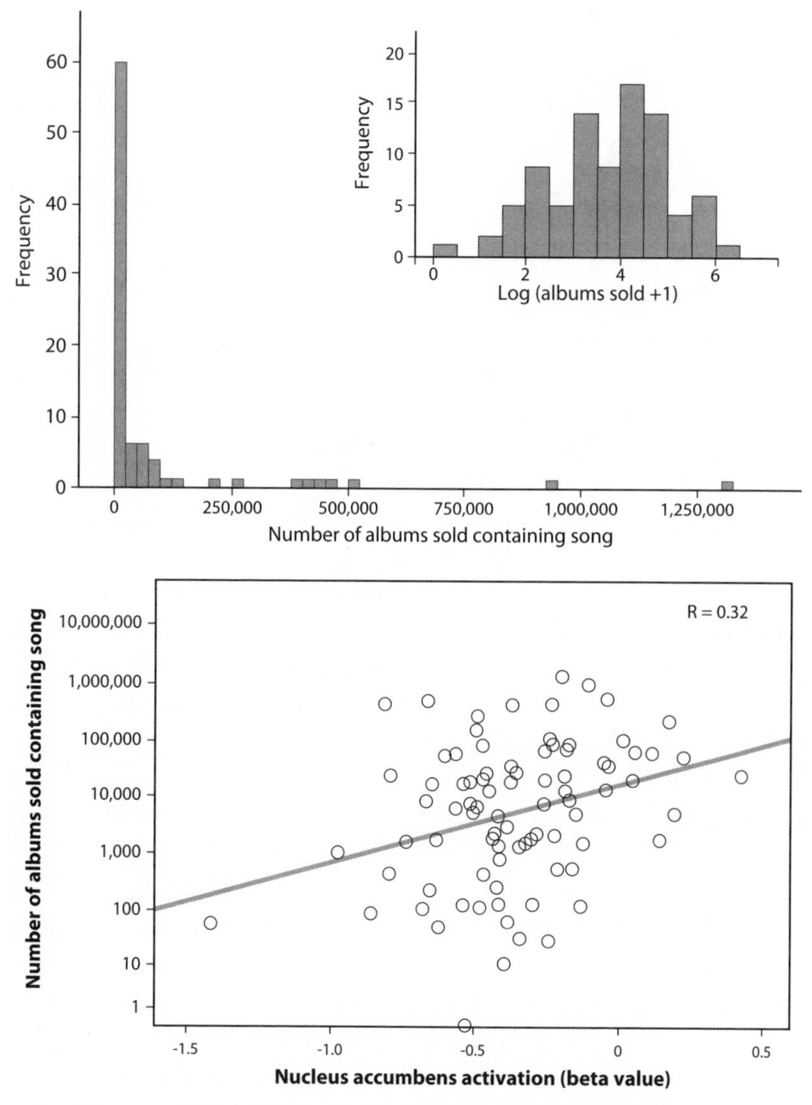

Figure 3.3 The number of albums sold (a) with the correlating fMRI activation of the nucleus accumbens (b).

Source: Berns and Moore 2011.

Evolving Perceptual and Motor Systems

Early investigators of the brain understood behavior as a chain of sensory/motor reflexes without cognitions. The brain, we now know, generates behavior by endogenous mechanisms that are not simply a chain of sensory/motor reflexes. A number of behavioral functions are fixed and demonstrate little flexibility, whereas others are more labile.[48] Everyday life is rife with the development of new motor habits.

The development of motor control is one of the fundamental events in the nervous system, and the development of muscles and nerves reflects endogenous processes. Endogenous processes do not take place in a vacuum, but within environments.

Codified rhythmic sequences run through music, neurally coded and culturally linked.[49] The neural substrates include sequencing of music. Timing in music is one essential feature of the sequencing of intervals. One region linked to tracking musical structure is the posterior region of the medial prefrontal cortex.[50] This region is tied to many regions of the cortex, including subcortical regions such as the basal ganglia.

The basal ganglia, along with the motor cortex and to some extent the cerebellum, are central to the regulation of motor control, not necessarily movement per se, but the "serial order of behavior," and to musical sensibility.[51] The basal ganglia are localized under the cerebral cortex and are composed of the striatum, which consists of the caudate and the putamen, in addition to the globus pallidus.[52] These regions are reciprocally connected to the frontal cortex and other cortical regions, as well as to other sites that underlie the organization of action.[53]

The anatomical connections between the frontal cortex and basal ganglia underlie the organization of action; central generators are a vital source in the brain's generation of action.[54] The caudate nucleus and the nucleus accumbens, part of the basal ganglia, are major areas in the organization of action. The nucleus accumbens is part of the circuitry by which motivational interceptive signals (desired goals that underlie behavior) from the hypothalamus and amygdala are transduced into motor output and behavior.[55]

It is apparent that a good deal of the nervous system is devoted to sensorimotor integration, but these sensorimotor systems are also integrated with cognitive information systems. Instead of envisioning the motor system as separate from cognitive systems as traditionally understood, the cognitive systems are embodied in the organization of motor systems (e.g., syntax).[56]

That certainly does not mean that cognitive systems are confined to motor systems; surely they are not. What is crucial to convey is that there is no separation of cognition from motor systems. I suggest that we don't view cognition and the organization of action (e.g., its initiation, basic programming, and execution) as being opposites. Instead, we should consider these two systems inextricably linked; no one piece of the organization of action in music is non-cognitive.

Basal Ganglia and the Organization of Action

From an evolutionary point of view, including behavioral adaptation, the organization of thought is embedded in the organization of action.

For instance, syntactical competence actually appears to be a form of motor control. In his article on the serial order of behavior, Lashley extended the language of syntax to basic motor control, opening the context for the cognitive presence within basic motor function. Lashley suggested that syntactical organization underlies the organization of diverse behaviors, not just language.[57]

One example of syntactical organization, in addition to syntactical organization in music, is simple behavioral sequences. The behaviors of several species have been studied for characteristic patterns of behavior. These behaviors (e.g., grooming) are species-specific and rule-generated. The behaviors have been studied under diverse conditions. Interestingly, one can exaggerate the movement sequences and decrease expression of the sequences under conditions of either too much or too little dopaminergic contribution.[58]

This serial coding of the behavioral expression is dependent on neo-striatal neurons, in which dopamine is an important neurotransmitter for normal behavioral expression.[59] Both too little and too much dopamine reduces competence and options. In music, as in other forms of human expression, the effects would be apparent. Too much would be an exaggeration and distract from musical sensibility, and too little would lead to defects in following the rhythms of the music.

Regions of the basal ganglia are linked not to the production of movements per se, but rather to the order of the sequences generating the organized sets of behaviors. As previously stated, dopamine plays an important role in the organization of thought and the sequencing of behavior. This organization is key to our ability to produce music.[60]

The basal ganglia are involved in a variety of learned behaviors. We know that, in diverse species, action sequences are represented in diverse regions of the brain, including the prefrontal cortex. We also know that parts of the basal ganglia, along with Broca's area and other regions of the frontal/motor cortex, underlie basic syntactical processes, including musical syntactical processes.[61]

The point I would like to emphasize is that the basal ganglia are involved in a wide range of cognitive functions, including that of music. Because of high concentrations of dopaminergic innervations, the basal ganglia are also linked to the motor and cognitive impairments of Parkinson's disease.

One emerging function of regions of the basal ganglia (e.g., striatum) is the prediction of rewards. For example, in human studies, as indicated, one set of dopamine transmissions in the striatum has been linked to the prediction of monetary rewards.[62] In fact, different parts of the basal ganglia have been linked to different phases in the learning of events. One study in humans using fMRI to measure brain activity has found that the ventral striatum is particularly linked to prediction of reward,[63] and dopamine neurons within this region are consistently tied to the learning of associations and the prediction of rewards.[64] For instance, the expectation of monetary rewards activated part of the nucleus accumbens.

Indeed, dopamine is released in regions of the brain during expectations of music, and while hearing music, it is released in regions of the brain that are critically involved in the organization of action, namely the basal ganglia and in particular the nucleus accumbens. In a study using positron emission tomography to measure dopamine release, parts of the caudate region of the brain were linked to the anticipation and, more strongly, the experience of music.[65]

Data from Parkinsonian patients and brain imaging studies also link the basal ganglia to rhythm. Indeed, one view is that the basal ganglia are essential as internal generators of rhythm.[66] Rhythmicity is an active event to which cognitive/motor coordination and integration are essential. Various forms of syntactical defects, including rhythmic impairment, have been linked to basal ganglia devolution.[67]

Oliver Sacks noted, in describing one of his Parkinsonian patients, that she had become "graceless" and that "she had lost her former naturalness and musicality of movement."[68] Yet, other observations suggest that some individuals suffering from motor control conditions actually perform much better under the influence of music (i.e., can sing when they can't talk, can dance when they can't pick up a spoon, etc.). In these cases, it appears that what is lost is the control of movement—what Hughlings Jackson, the great nineteenth-century neurologist, emphasized in the devolution of function, particularly control.[69] The surge in dopamine experienced by people involved in music perhaps temporarily affects their levels of control.

As I indicated earlier, motor systems underlie song and music, from the larynx to different regions of the central nervous system. I reiterate, there is no separation between motor and cognitive systems; all motor systems are embedded in information processing and cognitive appraisal, reflexive or not.[70]

Indeed, the basal ganglia are linked to reasoning about statistical probability, a rather advanced cognitive capacity. Diverse evidence links this region to rhythm generators, essential for movement, goal-directed behaviors, and music. fMRI studies measuring brain activation indicate changes in response to expected cadence perception.[71]

Music therapy has been shown to be beneficial for the treatment of brain-injured and Parkinsonian patients.[72] Perhaps a natural sychronization found in non-humans is preserved in devolution of function in these patients.[73] More generally, the therapeutic efficacy of music has been suggested across a number of neurological disorders (e.g., Tourette's syndrome, stroke).[74] Indeed, music therapy has proven efficacious in a number of contexts, including diverse forms of normal child development,[75] the treatment of autism, amelioration of social anxiety, and a reduction of cortisol secretion perhaps associated with social anxiety.[76]

Rhythm, like the melodic tunes in most folk songs,[77] seems to be something that almost anyone can easily respond to. It shows how action and perception are linked, although this is disputed by some who study tone deafness in humans.[78] Tone deafness seems separate from one's ability to perceive rhythm. We all probably know people who can't sing a note but can still sway to the beat of drums. Nonetheless, fMRI studies show that the basal ganglia are active when listening to diverse forms of rhythms and beats.[79]

Entrainment to sound is a feature of the inherent rhythmic activity of perhaps basal ganglia function, something vital in the organization of social action: being with others, dancing with others, or singing with others. The musical cognitive/motor system is entrained to temporal patterns, facilitating diverse memories and cephalic function. Interestingly, early musical training can also encourage a wide range of motor expressions, and music can even engage movement in Parkinsonian patients when they are "frozen."[80] The neurotransmitter is depleted by the impaired activation, by the reduced dopamine numbers, and the endless compensatory responses in the remaining dopamine fields.

The synchronization of bodily movements is a core feature in animal adaptation.[81] Amazingly, experimental variation of musical rhythms has shown synchronization of movements; it has been demonstrated that variation of musical beats when presented to the cockatoo bird results in changed rhythmic movement to match the musical beat.

Not surprisingly, core social coordination, rooted in movements common in coordination with one another, is also involved in predictive tracking. It is shaped and regulated by the basal ganglia (figure 3.4).

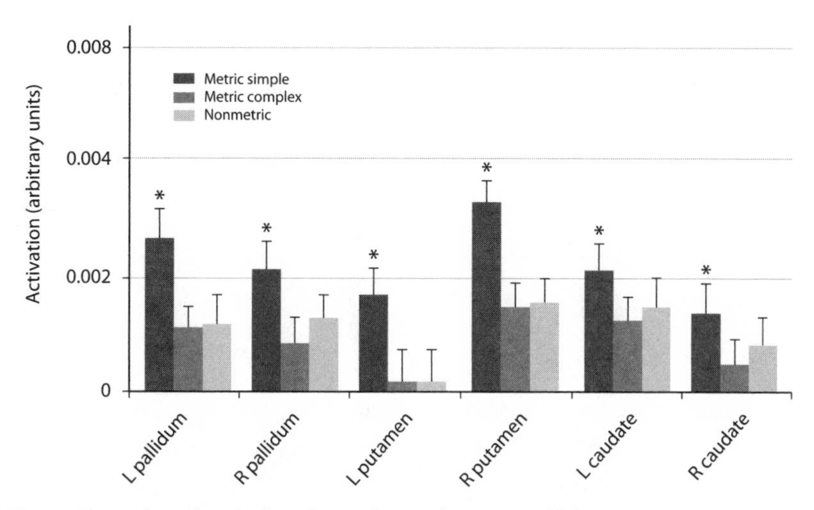

Figure 3.4 Left and right basal ganglia nuclei (i.e., pallidum, putamen, caudate) activation during various rhythm conditions in healthy volunteers.
Source: Grahn 2009; Grahn and Brett 2007.

When the basal ganglia are impacted by depletion of dopamine in Parkinsonian patients, one result, not surprisingly, is less competence in discerning differences in rhythms (figure 3.5). The basal ganglia are normally more active in fMRI studies when the beat is perceived as internally generated, as opposed to externally generated. Both are compromised by depletion of dopamine, as in Parkinsonian patients.[82]

Providing dopamine to Parkinsonian patients helps restore musical and rhythmic sensibilities. Moreover, Sacks noted that for his Parkinsonian patients "music indeed resists all attempts at hurrying or slowing and imposes its own tempo. I saw this recently at a recital by the eminent (and now Parkinsonian) composer and conductor Lukas Foss. He rocketed almost uncontrollably to the piano, but once there, played a Chopin nocturne with exquisite control and timing and grace—only to festinate once again as soon as the music ended."[83] The primordial effect of music on capability is profound and recruits diverse regions of the brain, and it does so because it is inherently tied to pleasure and to approach-related behaviors.

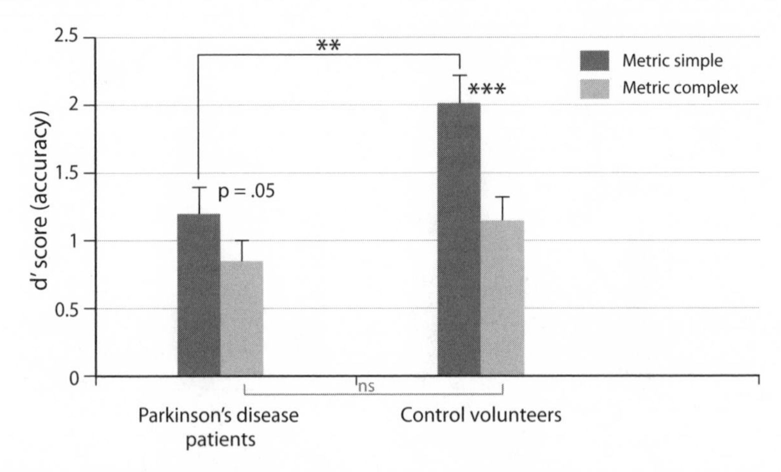

Figure 3.5 Decreased ability of Parkinsonian patients to discriminate changes in rhythm.

Source: Grahn and Brett 2009.

Procedural and Declarative Cognitive/Memory Systems

Memory is not a singular function; there are many memory systems.[84] Procedural memory and declarative memory are the two types of memory systems that we will consider for now. One system has been linked to procedures, rules, and syntax (skills); the other to semantic memory.

Perhaps mistakenly, declarative statements and memory (conscious and rule-governed processes that the individual is aware of) have been associated with cognitive procedures, whereas procedural memory (such as knowing how to play a flute "by heart") is construed as motor and non-cognitive.[85] I think this is a misleading way to characterize the differences between the two kinds of memory, for both reflect cognitive or information-processing systems. Distinctions like procedural and declarative memory have less to do with whether something is cognitive and more to do with whether something is habitual or new. Both memory types figure importantly in musical sensibility.

Memory tasks that require distinguishing what Gilbert Ryle called "knowing how" from "knowing that" are not assessing the difference between something that is cognitive and something that is not, but rather something that is automatic and well-rehearsed versus something that is neither automatic nor well-rehearsed.[86] Once something is automatic, perhaps it can be relegated to regions of the brain such as the striatum, regions that we know in mammals underlie motor control, developed habits of action, language, and music. We also know that regions of the frontal cortex that underlie complex behavioral function can be automatic. Indeed, many features of musical practice and performance are automatic but also endlessly cognitive.

We also know now that regions of the frontal cortex and striatum underlie the organization of syntactical structure. For example, the detection of verb displacements in a sentence activates brain regions such as Broca's area and regions of the basal ganglia, as demonstrated through the use of brain imaging techniques.[87] These sorts of data suggest that any distinction between motor and cognitive structure is very dubious.

One distinction between procedural and declarative memory, however, is useful in understanding language through deficits in linguistic expression following neurological damage. This is the distinction between systems that control established motor skills, including the syntactical features of language systems, from the larger lexical or semantic systems.[88] This distinction was first based on studies of neurological damage and then later on a theory of language acquisition.[89] Moreover, studies indicate a dissociation between syntactical rules for music, like language, from the diverse forms of memory that aid language or musical sensibility, and larger semantic and cognitive capabilities.[90]

Central dopamine is related to the syntactical side of language and, perhaps, to music and movement.[91] Regions of the brain that underlie language and music interact at all levels and experiences, demonstrating diverse effects on both forms of human expression.

Musicians have a greater ability to perceive violations in foreign languages than do non-musicians. Learning about music can contribute to the acquisition of both one's own and other languages.[92]

Musicians are more able to detect violations of the structure of the language than the more semiotic components. Moreover, there are overlapping neural networks for music, language, and song. The interactions between phonology and melody, between vowels and consonants, are endless. Both musical and linguistic processing impact perception of songs.[93] Speakers of tone-dependent languages like Chinese tend to have elevated musical pitch sensibilities.[94]

Indeed, investigators have outlined a diverse cortical network that bridges Wernicke's and Broca's areas in musical sensibility. These regions of the brain had been previously tied more to semantics than to syntax.[95] As I will continue to suggest, there is no one "music area" in the brain and a great many neural regions are active in all aspects of musical sensibility. Instead, Broca's and Wernicke's areas, so integral to language, are, in addition to many other regions, critical to music.

Broca's area has been consistently connected to the "motor" components of language,[96] in addition to a broad array of information-processing systems including musical syntax.[97] In fMRI studies measuring brain activation, for instance, activation of Broca's area is related to visual-spatial cognition in musicians.[98] There is some evidence of increased gray matter density in Broca's area in orchestra musicians (as compared to non-musicians).[99] Perhaps the general link between language, action and music is reflected in Broca's area activation.[100]

There does appear to be evidence for the distinction between regions of the brain such as the frontal cortex (including Broca's area) and the basal ganglia, that are important more for organizing the motor/cognitive functions, and regions of the brain such as the ventral temporal lobe that are more connected to cognitive/semantic or lexical organization.[101] In spite of these distinctions, both are linked to, and underlie, musical sensibility as a whole.[102]

These findings are consistent with the fact that the basal ganglia, in addition to the frontal cortex, have been linked to syntactical organization for human language. Parkinsonian patients have greater impairments with the syntactical features of language than with its lexical components.[103] That is, Parkinson's patients in whom dementia is not an issue may have greater difficulty with parsing a verb correctly than with knowing what the verb signifies, thus reinforcing the distinction between the procedural and declarative features of memory as they pertain to language, and perhaps as they pertain to song and music.

Parkinsonian patients retain the semantics, but are compromised on the syntax of language, and perhaps the same holds true for music. Maybe it would be more coherent if, instead of referring to cognitive and motor systems when discussing, for example, the basal ganglia and frontal cortex involvement in parsing sentences or generating serial grooming sequences, we instead referred to the diverse cognitive systems that underlie motor sequences in behavior.

Further Considerations of the Neural Regions Responsive to Novel and Unusual Events

Diverse cognitive systems underlie musical experiences, running from emotional appraisal systems, to attentional systems, to syntactical and tonal processes, to rhythmic generators.[104] The forms of music that we learn and participate in set the background for the familiar and the unfamiliar, the novel and the unusual.

Discrepancy detectors, for which there must be many variants of human experience, reflect different systems in the brain and perhaps underlie some forms of aesthetic judgment.[105] The recognition of a discrepant event is a broad-based heuristic that pervades the organization of decisions and action. Moreover, syntactical processing and detection of discrepant syntactical events are picked up early on by regions of the brain that include the frontal and temporal cortices.[106]

One region of the brain, the amygdala, is linked to the detection of a number of discrepant events, possibly including facial discrepant events.[107] The amygdala also underlies a number of positive and negative responses.[108] Fearful or shy children show greater amygdala response to social stimuli and more generally to discrepant events.[109]

Interestingly, our understanding of the reiterations of amygdala function has evolved from olfaction and visceral regulation to social interpretation and perception, the magnification of events, reproductive behaviors, attachment behaviors, fear-related systems, the regulation of appetites, uncertainty, expectancy, and attentional responses.[110] Nonetheless, one core characteristic associated with amygdala function includes the detection of discrepant events and the readiness to perceive an event regardless of whether the event is dangerous or not.

The amygdala is activated to discrepant, unexpected musical events as well.[111] In one study, subjects were presented with an irregular array of chord sequences. fMRI of the amygdala showed greater activation to the unexpected musical syntactical sequence. Moreover, dopaminergic neurons within the amygdala are elevated during problem solving, and they may also underlie the perception of reward and the detection of novelty.[112]

These results fit within the larger category of results linking the appraisal of unfamiliar events. Musical memory and expectation are of course linked to the larger culture in which we are embedded, as is the perception of tone.[113] Tone is important for grasping the meaning of sound.[114]

In patients with dorsal lateral prefrontal damage, in which there is a large amount of dopamine innervation, impairment in responding to novel and unusual events is observed.[115] Decrements in responsiveness to the world reflect, in part, sensory neglect, reduced attentional abilities, and a lack of orientation to novel events. Regions of the brain such as the cingulate cortex, a massive and complex structure that can include the amygdala and the frontal cortex, are involved in both attentional and emotional mechanisms.[116]

Uncertainty follows from the recognition of a discrepancy, and when expectations begin to falter, what could formerly be relied upon no longer has the same weighted value. A search for a new solution emerges. In human experiments, the anterior cingulate and lateral orbitofrontal cortex are activated by uncertainty. When subjects are playing cards, for instance, and monetary awards are at stake, autonomic output is reflected in cortical activation.[117] The cortical regions, in addition to regions of the basal ganglia, are linked to a wide range of discrepant events and underlie the organization of action: to approach or avoid objects. Detecting musical notes that are unexpected is a rather general response.[118]

These events, again, may reflect the activation of dopamine innervation from the brain stem ventral tegmental dopamine neurons. The activation of this system (dopamine) may reflect in part a prediction error, a difference between expectation and outcome. With regard to dopaminergic neurons, we now know that they play a functional role in syntactical competence in humans, in musical competence, and more generally in aesthetic judgment.[119] In the context of listening to a piece of music, diverse regions that underlie dopamine activation should be activated by musical/syntactical discrepancy.[120]

Not surprisingly, the sequencing of music is related to activity in the frontal lobe and basal ganglia, and the anticipatory imagery of sound is similarly tied to the prefrontal and premotor regions.[121] Imagery of the familiar sound sequences and anticipated expectations activated these regions in some studies.

Conclusion

Categorical boundary conditions help set the stage for speech and music.[122] Cognitive motor systems are certainly involved, as Dewey understood and Meyer adumbrated, in the context of music in relationship to diverse cognitive motor systems.[123]

Discrepancy is just one feature in our experiences, but an important one, in the organization of action toward events in the world, including that of reward and musical events. Cognitive systems impact every level of rhythmic capacity in music and dance.

Neural transmitters, such as dopamine, play many roles in the organization of thought and action, and migrate to diverse end-organ regions within the brain and periphery. Neural systems reflect an evolution of cognitive speciation in part through the migration and utilization of dopamine in the brain and peripheral tissues that underlie thought, action, and musical expression. The motor regions are expanded in function amidst the selection of functional cephalic capability. Further access to cognitive/neural capacities that are expanded in use result in greater behavioral options; music is tied to the expansion of core features.[124] Moreover, the activation of dopaminergic systems under uncertainty is linked to reward mechanisms in the brain. Predictability of a stimulus has a direct impact on the human brain reward systems under a variety of experimental conditions.[125]

CHAPTER 4

Musical Expectations, Probability, and Aesthetics

John Dewey, in his ground-breaking book *The Quest for Certainty*, demythologized human knowing by highlighting real world problem solving within a framework of endless uncertainty. We seek to secure the stable and hold it while we grapple with endless uncertainty. We make our peace with the uncertain as we forage within inquiry and infuse " . . . objects of sense which are also objects which satisfy, reward and feed intelligence . . . through ideas that are experimental and operative."[1]

Grappling with uncertainty is a commonplace occurrence across diverse species. Music is lodged within the familiar to us and amid horizons of the less familiar as we explore new domains. Meyer understood and used the term "embodied meaning" for musical experiences with regard to the transactions of human beings in the context of musical meaning in our lives.[2]

Emotional response to music often comes in the familiar songs of the past. One of my favorite experiences was singing to the musical sensibility of Mrs. Ginny Armat, a cabaret singer in Washington during the 1940s. Her memory pervaded those songs. Her motor capability was compromised in her nineties, but she could still play, scaffolding by diverse memory props.

This chapter focuses on the human problem of grappling with the unexpected while sustaining the familiar, a core feature that underlies human experience and inquiry in general, but which is also essential, in part, to musical sensibilities. One neurochemical that is broadly linked to expectations and the organization of behavior, including that of music, is dopamine.

Expectations and the Aesthetics of the Everyday

Musical expectations permeate diverse forms of problem solving, and one of those is coping with "the shocks and instabilities, the conflicts and resolution"[3] that we experience every day.

The aesthetic sensibility is not simply a sensory pang, though sensory experiences are certainly part of it,[4] amidst diverse cognitive predilections.[5] Rather, a narrow view of the hedonic has difficulty taking into account grappling with the unexpected which lies at the heart of aesthetics, for "the tension and the struggle has its gatherings of energy."[6] It is this energy and this struggle with the unknown that is tied to inquiry, learning, and curiosity.[7] Music depends on the contrast of the familiar and its variation, something Meyer pointed out. There are many threads to Meyer's musical trajectory. As Robert Gjerdingen, a musicologist at Northwestern University, notes, "there was the humanistic historian of listening (Style and Music), the systematic analyst of musical structure (Explaining Music), the performance coach (The Rhythmic Structure of Music), the music critic (Music, The Arts and Ideas), and the music psychologist (Emotion and Meaning in Music)."[8]

Of course, in some contexts the repetition of the familiar is part of the experience of music. Indonesian gamelan bands depend on the repetition of an already-familiar musical motif with minor tone variations in some form of endless, repetitive, "eternal recurrence," to borrow a phrase from Nietzche.[9]

Variation at many levels of appraisal, however, is a feature of most musical experience. Both Mozart and Coltrane delighted in slight deviations in expectations and musical sensibilities pervading the familiar. John Coltrane's recording of the song, "Some of My Favorite Things," written by Oscar Hammerstein and Richard Rodgers for *The Sound of Music*, relies on the familiarity of the audience with the song to launch into magnificent variations, expected and unexpected, as he expanded into novel uncharted territory.

The play of ideas amidst the uncertainty of one's existential condition defines an aesthetic in which probability expectations are part of our everyday cognitive adaptation, in music or otherwise. The codification of well-worn habits and expectations is the grounding frame in which musical orientation takes place; it requires well-orchestrated habits, cognitively mediated in regions of the brain that underlie repetitive occurrences and their breakdown. Dewey suggests that "The road from a perceptible experience which is blind, obscure and fragmentary, meager in meaning, to objects of sense which are also objects which satisfy, reward and feed intelligence is through ideas that are experimental and operative."[10]

This pragmatic vision is embedded in a sense of music, and more generally in aesthetic and diverse forms of inquiry.[11] Adaptation is a core feature of musical sensibility; the biological systems are tied to auditory acuity and the prediction of auditory events.[12] Appraisal systems in musical pleasure and displeasure are embedded in the auditory system analysis and the larger connectivity in cephalic systems. Infused within this are diverse forms of learning, rich in semiotics.[13]

For pragmatists anchored to biology, such as Meyer, pragmatism is connected to aesthetic experiences inherent in everyday life and in shared experiences. In this view, expectations are pervasive within the organization of action, musical or otherwise.[14]

The musicologist David Huron, in his book *Sweet Anticipation*, outlines in a masterful manner the direct relationship of Meyer's view about expectation being endemic to musical sensibility, providing an expansive sense of the biological side of musical sensibility from the experiences of expectation in simple and complex music.[15]

We have a wide variety of musical expectations that range, as Huron notes, from pitch proximity to the direction of the music. He ties this organizational sensibility to "heuristic listening." Scale tones set up diverse expectations amidst surprises and "sweet anticipation." Indeed,

> . . . some surprises start right from the moment a work begins. When Igor Stravinsky began his "Rite of Spring" with a solo bassoon, he violated several well-known conventions in classical music. The vast majority of Western orchestral works do not begin with a solo. Moreover, the bassoon is one of the least likely orchestral instruments to perform by itself. Finally, Stravinsky placed the instrument at the very top of its range. In other words, Stravinsky began the "Rite of Spring" in a highly unorthodox (that is, improbable) way.[16]

That shock of unexpectedness mimics the way the season of spring itself surprises us after the cold of winter.

Expectations and variations are common themes for which musical structure sets the conditions for diverse musical expression. We create the same sort of variation on themes by naming places in unfamiliar regions after familiar ones as we explore the new (New Amsterdam and New York, for instance, as names for the settlements on Manhattan, first by the Dutch and then by the English), as well as by exploring new food sources while accompanying them with familiar condiments (think of the way children can be introduced to new foods by the addition of ketchup).[17] In music, we similarly explore the unfamiliar; perhaps it is another way in which we attempt to "tame chance."[18]

Surprise and tension underlie most of human experience, but they take on new and extraordinary forms in music.[19] Indeed, such events pervade the sense of musical experience.[20] Consider, for instance, the expectations we have concerning melodic intervals.[21] Cephalic capabilities underlie diverse forms of human expression, and predictability is a core feature of the brain's responses to rhythmic expectations.[22] The sense of musical expectations is embedded in appraisal systems, replete with emotions.[23]

Aesthetic judgment is ultimately a profound part of our life. We do not just view a masterfully rendered painting with awe. We value a person's character as a work of beauty, a beautiful soul; we enjoy the aesthetic contrast of orange carrots and green peas on our dinner plates; a majestic mountain glimpsed through a car window renders us speechless as it comes into view. Aesthetics is pervasive in everyday life. It is not just for the museum, parlor, or theater.

Aesthetic judgments draw one close, eliciting approach behaviors, but they also can repulse and make us withdraw from objects. John Dewey, in his book *Art as Experience*, stated romantically, "Because experience is the fulfillment of an organism in its struggles and achievements in a world of things, it is art in germ. Even in its rudimentary forms, it contains the promise of that delightful perception which is aesthetic experience."[24]

Art is thus the stuff of everyday life. Aesthetics is a profound part of our experience. It heightens and deepens our experience. It is not something that closes us off, but expands us beyond ourselves and toward others. Art is a vital source of communication and it is therefore perhaps not surprising that rudimentary forms of aesthetic appreciation, song for example, may be expressed in a variety of species, and are reflected in diverse cephalic structures.[25] Art is a feature of the human condition, and all cultures have music of some sort.[26]

Functional units that facilitate successful behavioral and reproductive strategies are recruited from a number of information processing systems, aesthetic and otherwise. They serve diverse purposes.

Aesthetics, Expectations, and Learning

Within learning theory, prediction is coupled with expectancies and their breakdown.[27] When expectations are thwarted, a broad array of learning occurs through new problem solving and search principles. This is close in scope to Peirce's view of inquiry and the development of new solutions to problems.[28] Of course, inquiry is more than this. Anthony Dickinson demonstrated the link between causal inference and prediction in a wide array of animals.[29] Peirce's view of inquiry and learning was prescient, for the variants of this view would capture learning theory through what became known as the Rescorla-Wagner model:[30]

$$\Delta V = \alpha \beta (\lambda - V).$$

The Rescorla-Wagner model depicts the associative strengths of stimuli and how discrepancies from expectations are resolved. An association, and thereby learning, occurs by the strength of the predictions that are being developed. The model then is not simply a mathematical approach to neural science, but also an incorporation of a cognitive point of view. In the equation, V represents the current associative strength of the stimulus, while λ shows the maximum associative strength of the primary motivating event. The salience of conditioned and unconditioned stimuli is represented by α and β respectively. The predictability of the primary motivating event is shown in the $(\lambda - V)$ term. When the current and maximum associative strengths of the stimulus are equal, the conditioned stimulus fully predicts the reinforcer. However, when the term is positive (λ is greater than V), the associative strength increases and the conditioned stimulus does not fully predict the reinforcer—there is room for learning to occur. With increased associative strength, learning occurs, and in fact only occurs when the conditioned stimulus does not entirely predict the reinforcer. In contrast, a negative $(\lambda - V)$ term occurs when there is a loss of associative strength and the predicted reinforcer has failed (extinction).

General informational search and discrepancy mechanisms, such as the one outlined above, feature underlying musical sensibility. As Dewey understood it, we search for the stable amidst the precarious; we search for the predictive. I would express it a little differently: rather than the stable (homeostasis, or staying the same), we search for the viable so we can adapt to changing circumstances (allostasis).

An aesthetic of viability is functional amidst change, but that change is embedded in goals. As Meyer put it, "The very concept of an Avant Garde implies goal directed motion—the conquest of some new territory."[31] Meyer's emphasis, like Dewey's, is on stability and probably implicit in homeostasis. Allostasis, on the other hand, is cephalic in origin and directed toward the biology of change.[32]

Probability, Expectations, and Learning

Probability judgments to assess the condition of uncertainty are at the heart of human reasoning, and while it is a mistake to exaggerate their role in aesthetic judgment, one can assume that in some contexts, they do play a role. Since uncertainty is a basic feature of our existence, we have developed a variety of resources to cope with it. Moreover, as Meyer has noted, " . . . uncertainty is anathema to humankind" and "we devise ways of reducing uncertainty both in the out-there world and in our personal lives," but " . . . in the arts and other playful activities such as sports, games and gambling we actually relish and cultivate a considerable amount of uncertainty."[33]

John Sloboda, a professor of psychology and long time researcher in the psychology of music, noted in a paper entitled "Leonard Meyer: Embracing Uncertainty" that in Meyer's view "emotion is intimately linked to the cognitive processes of generating and testing predictions about future musical events."[34]

So, while we eschew uncertainty in our daily lives, we are also emotionally involved with, and "play" with, its reverberations in music and other aesthetic expressions. Probability and detection of discrepancies are a definitive part of problem solving and are also one part of aesthetic judgment. Many issues related to probability have been couched in terms of expectations. Indeed, syntactic predictions may have their basis in general issues about probability and human reasoning.

Considerable data link curiosity to drive states.[35] The literature on this emerged several decades ago during the height of drive reduction theory (drives meaning some sort of need, biological or otherwise, that motivates behaviors).[36] The more modern variant places a curiosity drive in the context of filling in some information: a gap in knowledge is noted, information is needed, and a drive to fill the gap emerges.[37] Variants of this perspective hearken back to Freud and others, who construed curiosity as a "thirst for knowledge," because curiosity increases our propensity to acquire new information.[38] What underlies musical sensibility, like other achievements, is a cephalic structure oriented toward change.

The sense of change is what underlies the appreciation of knowledge in music and adaptation; the sense of uncertainty is part of the interplay of the regularity of the expected with a deviation, however slight, that keeps our interest. Such is the structure of musical expectation, amid style and form. Staying the same, alternatively, is surely not a feature of many musical forms, although variation within set parameters is a recurrent musical feature.

Discrepancy, Uncertainty, and Change

In ordinary everyday activities, from the pursuit of scientific knowledge to aesthetic appreciation, the detection of a discrepant event motivates behavior.[39] The behavior serves, in part, to quell the sense of discrepancy; many authors have noted the sense of achievement experienced in reaching cognitive equilibrium.[40] But perhaps it is not equilibrium, but adapting to change amidst the familiar: something called "allostasis."

A "curiosity drive" was invoked to explain why monkeys would press an operant to look at something or to observe an event.[41] An ethological view, one not shared by earlier drive reduction behaviorists, is that animals are prepared to learn, to notice events, and to store that information for later use.[42] There is pleasure in the knowing process, whether related to drives or not.

Moreover, this curiosity drive is less in the shadow of internal processes and more construed by interaction with environments; not a mind cut off from the world, but one interacting with environmental factors. The behaviors expressed reflect the options available in the environment, with one core feature being novelty of stimuli (figure 4.1).[43]

The curiosity drive seems to be related to the exploration of novel objects and is essential for understanding the world to which one is trying to adapt.[44] The drive to acquire a new habit can be shown through behavioral patterns that are aligned with the breakdown of a previous habit.[45]

There can be no doubt that when information is unusual, there is a strong desire to acquire it. A broad-based response to discrepancy is an important behavioral adaptation, underlying aesthetic sensibility; curiosity can be pleasurable. A sense of pleasure in music is related to novelty.[46]

In the discrepancy model, disruptions of expected events result in the recruitment of a greater number of behaviors that might re-

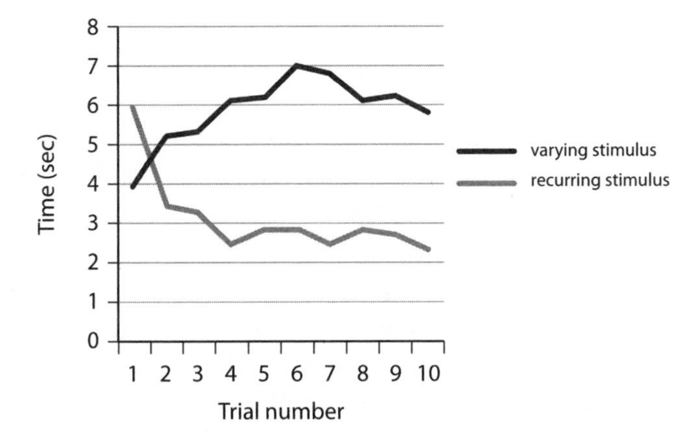

Figure 4.1 Responses to varying versus recurring stimuli; the more novel a stimulus, the greater the time required for the elicitation of a response.
Source: Berlyne 1958.

flect learning. Functionalist and cognitivist views place the behavior patterns that are generated by a central state of the brain in the context of acquiring information, which has been consistently linked to the context of musical sensibility.[47]

Information acquisition and control is a strong desire. Aversiveness to ambiguity, for example, is a real property of our decision making. Ambiguity can breed indecision and attempts to preserve the status quo. Interestingly, George Loewenstein,[48] noting a very interesting observation that James made, expressed that we are prepared to recognize discrepancy, and we then search to fill in the gaps.[49] It remains clear, however, that resolving uncertainty is a major motivator of behavior.[50] Perhaps this might make us eager, sometimes, to hear the end of a song that defies our expectations.

Interestingly, information deprivation is construed as a cognitive deprivation. A consequent hunger to fill this gap occurs. Within the bounds of reason, some forms of curiosity may reflect this search for cognitive equilibrium. Curiosity emerges when one's knowledge and reference points are recognized to be inadequate.[51] Human experiences in decision making suggest that human choice is based on informational variables that reduce uncertainty. Drawing on animal experiments, motivation increases at the point of resolving uncertainty, of finding a solution.[52]

In one experiment, for example, subjects were shown parts of a human body in a visual array presented to evoke curiosity. Different groups of subjects were shown different numbers of body parts such as hands, feet, a torso, and so on. Subjects were asked to predict the age of the person depicted in the collection of body parts. They were also asked how curious they were to find out the age. Loewenstein et al. predicted that those subjects shown a greater number of body parts and thus afforded a greater opportunity for visualizing the body as a whole would be more curious about the age of the depicted person (figure 4.2).[53] In other words, the propensity to fill in the perceived information gap should be stronger in those shown more information about the possible age of the subject, and this should be correlated with the subjects' self-reports of curiosity. The experimenters found corroborative evidence for these expectations.

The emphasis on *visceral input* is vital to curiosity, learning, and inquiry. Interestingly, in the common phrase "to pique one's curiosity," the word pique means "to arouse, provoke," but the word can also mean "to cause to feel vexation or resentment." The gap of information, both relative and absolute, leads to a possible understanding of curiosity and the joys that one might find in the search for and attainment of information. The relationship of the information gap with curiosity has been articulated through the use of information theory's entropy coefficient:

$$n - \Sigma\, p_i \log_2 p_i$$

$$i = 1.$$

In this equation, n represents the total possible choices or outcomes of information, whereas p_i is the assessed probability that a particular choice will occur.[54] As knowledge concerning each choice increases, the probabilities of each become more varied and exact. The equation is useful, but not necessarily exact in quantifying multidimensional information in a unidimensional manner. In terms of the information gap and curiosity levels, several en-

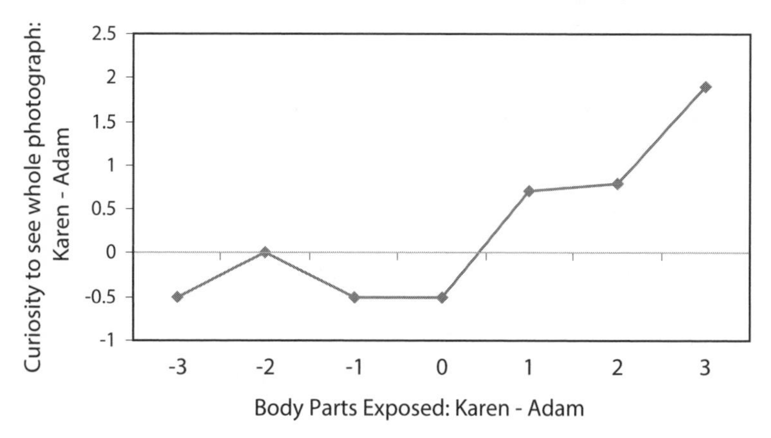

Figure 4.2 Curiosity and exploration.
Source: Loewenstein 1996, unpublished.

tropy measures are necessary: the individual's current situation, the informational goal of the individual, and a situational level of ignorance are all factors. The absolute magnitude would therefore be the informational goal minus the current situation. The relative magnitude would be the absolute magnitude divided by the difference of the informational goal and level of total ignorance.

The major point is that people tend to rely on both the relative and absolute magnitudes of the information gap in order to close the gap.[55] We look, in music, certainly as Meyer suggested, and elsewhere, to fill in the gaps; we hear the gaps. Some forms of curiosity within the musical experience are related to visceral factors, and there is no doubt that curiosity is linked to frameworks, or to a storehouse of knowledge. One issue is what kinds of gaps we might find in music and how this sense of curiosity is stimulated when listening to or producing music.

Our urge toward exploration is wider than any narrow view of expectations and violations for motivated musical behaviors. It is a fundamental part of the structure that we impose, the expectations that pervade our experience, the codified habits instantiated in cephalic organization that inhabit the perception and motor expression in music. But variation is a common theme. The same holds for music: violations of expectations will not totally explain why new music emerges.[56]

We forge coherence by relying on the familiar amidst the unfamiliar, and it is this relationship that permeates much of our experience in music, and indeed in most avenues of human expression. Meyer understood that melodies are often acting on more than one level, in which there is the familiar and the unexpected. We are anchored to frames of reference in which there are expectations, detection of deviations or gaps, and then exploration. Continuity, building on the insights of Peirce, is a primary feature of expectations situated in a cultural and contextual milieu, within a "cultural style."[57]

Expectations are a core feature of cephalic organization. Repetition is essential in habit formation, and small deviations underlie responses to even the smallest novelties.[58] These factors operate in music as well as in daily life.[59] Expectations underlie a pattern itself, and are indeed part of the pattern. A common arrangement in Western music (and also Western humor) is the repetition of two similar phrases followed by a dissimilar event, the A-A-B pattern.[60] So, a break in expectation is perhaps part of sustaining coherence.[61] The results from one study are depicted in table 4.1.

Of course the tension in a piece linked to expectation and variation from expectation is an underlying factor; tension and release, buildup and reduction, are core features of the musical aesthetic. One theory links these interactions of affective tone to emotional expression.[62] For instance, melodic expectations are a continuum in our musical sensibility, and also appear to be tied to information processing, the sound of music (the tension and release).

We are curious for a lot of reasons, and there is no doubt that one of these reasons is informational discrepancy in an event. However, the feeling of the information discrepancy may not necessarily be aversive; deprivation is aversive by nature, but curiosity is not; perhaps it is the inherency of risk, before it was tamed and became safe, in the unknown that makes us desire to understand. Maybe it is really a fear of the unknown that drives our curiosity . . . yet somehow that quest has taken on the appearance of pleasure . . . like a roller coaster ride or a scary movie. As Loewenstein rightly

TABLE 4.1

Distribution of Melodic Patterns Across Various Genres

	AB		AAB		AAAB	
	N	%	N	%	N	%
Western/art	29	35.3	44	53.7	9	11
Jazz/popular	23	21.1	73	67	13	11.9
Jokes 200	31	20.5	109	72.2	11	7.3

Source: Rozin and Rozin 2006.

acknowledges, we are motivated by more than simply appetitive reasons to do what we do (e.g., climb a mountain).[63]

Now, curiosity can reasonably be recruited to serve the drive to fill in some information in a setting in which the discrepancy is unsettling or viscerally displeasing.[64] It is a stretch, however, to identify curiosity only as a drive, and an aversive one at that. Cognitive equilibrium is a real metaphorical extension of bodily notions of homeostasis.

Peirce also linked "play" or "musement" with the induction of ideas, like a lot of us. An important adaptation is the fact that we can really enjoy gently musing about things. Sometimes interesting ideas emerge at these times. New ideas are hard to find in any context, so why exclude the playful part of the mind? Does it seem less serious because it is not labored, fought over, or an overcoming of adversity? The play of ideas is something sweet and precious to us, a luxury to be savored and enjoyed, and, where possible, extended. It is no less real or important because an idea is constructed with a sense of play. However, it seems reasonable that we currently take pleasure from curiosity and that we are often rewarded for it, but that its original purpose was to help keep us safe.

Musical Expectations, Discrepancy, and Aesthetic Judgment

Cognition is often construed as detached. Since emotions are anything but detached, they are considered noncognitive by many investigators. Aesthetic experience is surely up close and personal, and therefore one can understand why many investigators construe the aesthetic experience as not "exclusively cognitive."[65] I, on the other hand, don't know what part of aesthetics is not linked to information processing systems in the brain. In other words, aesthetic appreciation is replete with information processing; it is not as if one side of us is doing the thinking and another is only appreciating.[66] In fact, cognitive systems underlie all aspects of music and aesthetic sensibility.[67] It is a pernicious if deeply ingrained separation to say that emotions and cognitions are tangibly different.

David Temperley, a professor at the Eastman School of Music, has written some informatively integrated probability models (such as Bayes' theorem). Bayes' theorem is a fundamental theorem of probability that states that, for any two events A and B, the probability of A given B can be computed from the probability of B given A, as well as the overall probabilities (known as the "prior probabilities") of A and B.

Temperley has rather nicely linked Bayes' theorem to an understanding of statistical expectations and their violation in music. Statistical reasoning, like many forms of cognitive adaptation, is not conscious. Temperley, writing about Meyer in a paper on Bayes' theorem in music, put it this way:

> More than forty years ago, Leonard B. Meyer remarked on the fundamental link between musical style, perception, and probability: "Once a musical style has become part of the habit responses of composers, performers, and practiced listeners it may be regarded as a complex system of probabilities. . . . Out of such internalized probability systems arise the expectations—the tendencies—upon which musical meaning is built. . . . The probability relationships embodied in a particular musical style together with the various modes of mental behavior involved in the perception and understanding of the materials of the style constitute the *norms* of the style.
>
> . . . To me (and I believe to many others who have read them), these words ring profoundly true; they seem to capture something essential about the nature of music and musical communication. Building on

> these ideas—towards an understanding of how probabilities shape musical style and perception—would seem to be a natural enterprise for music cognition and music theory. Perhaps surprisingly, the probabilistic approach to musical modeling has not been very widely explored.
>
> . . . The Bayesian approach, I will argue, opens the door to a more musically sophisticated investigation of the probabilistic aspect of music than has been possible before. The Bayesian approach is inherently concerned with *structure*, and the relationship between structure and surface: the way structures constrain surfaces in composition, and the way surfaces convey structures in perception.[68]

Temperley's hypothesis is highly speculative, but interesting. His main point is that the systems that are implicated in the organization of action and perception in everyday life events (and perhaps something like Bayes' theorem, which is one mechanism among others used to track events) also underlies musical sensibility; the modes are all about probability in melodic perception, for instance.[69] Importantly, the structural systems are inherent to the perception and action in the appraisals. But, for many of us there is no separation; there are just forms of information systems, of which emotional appraisals or communicative function are just one among others.

Expectation, Musical Meaning, and the Emotions

In the context of music, the melodic structure is linked to implications or expectations and their realization.[70] While the experience may not reflect the expectation, the way of explaining the behaviors and understanding the mechanisms do. Acknowledging Meyer, musicologists have long noted the inherent gestalt-like properties of music.

Moreover, something inherent in Darwin, who did not acknowledge a separation of cognitive systems from emotional systems, is of course the view of Meyer and certainly Dewey: that emotional responses are embedded in diverse forms of information processing

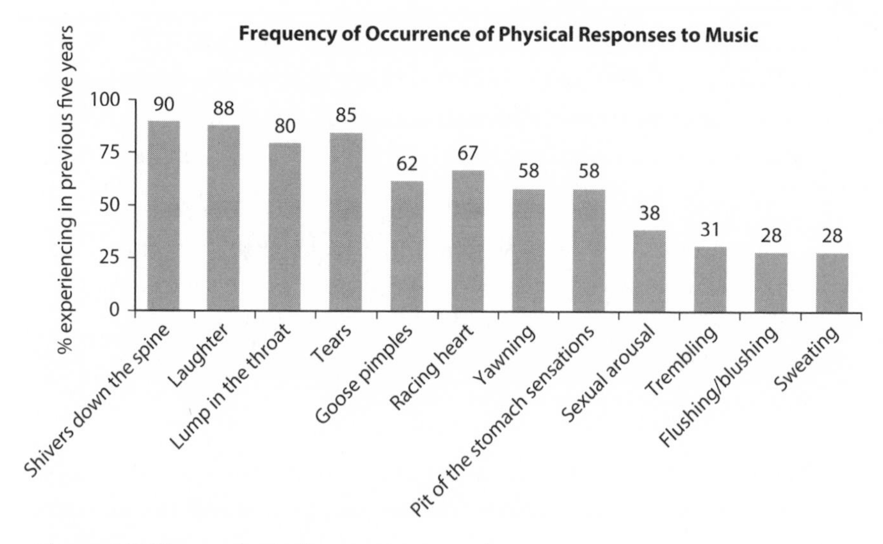

Figure 4.3 Diverse physical responses to music.
Source: Sloboda 1991.

cephalic systems.[71] Indeed, Sloboda, a psychologist who studies musical sensibility, has commented on Meyer's work regarding emotion: "Emotion is intimately tied to the cognitive processes of generating and testing predictions about future musical events."[72]

Sloboda comments further on Meyer, noting that emotional responses and changes to the music reflect "implicative relationships that Meyer proposed."[73] Violations of expectations are one reliable feature linked to emotions, and, while Meyer had doubts about the accuracy of the emotions in music, physical embodiment is embedded in these expectations and their violation in musical appreciation.

One feature inherent in some forms of music is tension,[74] a conflict about expectations inherent in the psychophysics of harmonic combinations. Sloboda depicted a set of common physical responses to music (see figure 4.3).[75] The responses are quite varied, not surprisingly, and they capture a wide range of human emotions.[76]

Patrik Juslin, a psychologist of music from Sweden, adapted the approach of twentieth-century experimental psychologist, Egon Brunswik, in tying music to emotional communication (table 4.2).[77] Brunswik addressed the probabilistic nature of inferences in adaptation. These are the kinds of adaptive responses that also run through music and the emotional communication inherent in the expression of music.

Some types of music are tied to core sets of emotions. This is very much in the tradition of Darwin and James, who both believed that emotions bear functional relationships that enhance

TABLE 4.2

Brunswik's Approach to Emotional Communication in Musical Performance

Emotion	Cues
Happiness	Fast tempo, high sound level, bright spectrum, staccato articulation, fast tone onsets, moderate variations in timing, sharp durational contracts (i.e., between "long" and "short" notes)
Sadness	Legato articulation, low sound level, slow tempo, soft spectrum, slow tone onsets, slow and deep vibrato, soft durational contrasts, relatively large deviations in timing, final ritardando
Anger	High sound level, sharp spectrum, fast tempo, staccato articulation, abrupt tone onsets, mostly sharp durational contrasts, heavy vibrato, spectral noise, no final ritardando
Tenderness	Legato articulation, medium sound level, slow onsets, fairly slow tempo, intense vibrato, fairly large deviations in timing, soft spectrum, mostly soft durational contrasts
Fear	Extremely staccato articulation, very low sound level, low (average) tempo, soft spectrum, large tempo variation, pauses between phrases, much dynamic variation, intense and irregular vibrato

Source: Juslin 2001.

cephalic capability. This is their link between two core features in us, namely musical sensibility and a conception of probabilistic reasoning with diverse forms of expectations; in the same manner, Juslin, following Brunswik, links functionalism, adaptation, and probability in responses.[78]

In more detailed studies, the greater the deviation from expectations about the music's harmonic characteristics, the greater the emotional response (figure 4.4).[79] That is, cephalic systems come prepared to recognize harmonic relationships, syntactically organized in the diverse semantic meanings expressed in music; deviations result in greater and more diverse cephalic responses.

The attention here becomes focused when encountering something that stands out as discrepant. Our curiosity is piqued; beholding an aesthetically pleasing or offensive object, our interest is aroused, our sensibility offended.[80] Expectancy in information processing in the brain permeates and results in what Karl Lashley called "the serial order of behavior."[81]

Lashley, among others, has suggested that, "Input is never into a quiescent or static system, but always into a system which is already actively excited and organized," and "only when we can state the general characteristics of this background of excitation, can we understand the effects of a given input."[82] We know something about the central control of motor or syntax programs. The serial organization of appetitive and consummatory behaviors reflects the central organization of the brain. They no doubt underlie some of our aesthetic judgments.

John Dewey also held this view of neural preparedness, and is closely associated with theories in which behavior toward aesthetic objects reveals both appetitive and consummatory experiences in which stability is sought while precariousness is restrained.[83] All of us, in the search for stable and secure relationships, experience both the appetitive, or the search mode, and the consummatory outcome.[84]

Emotion, in this view, only becomes salient when a habit, a way of behaving that has been successful, is no longer effective. Of course, this renders emotions only in the context of conflict, and I would suggest that emotions are not just about conflict, its ex-

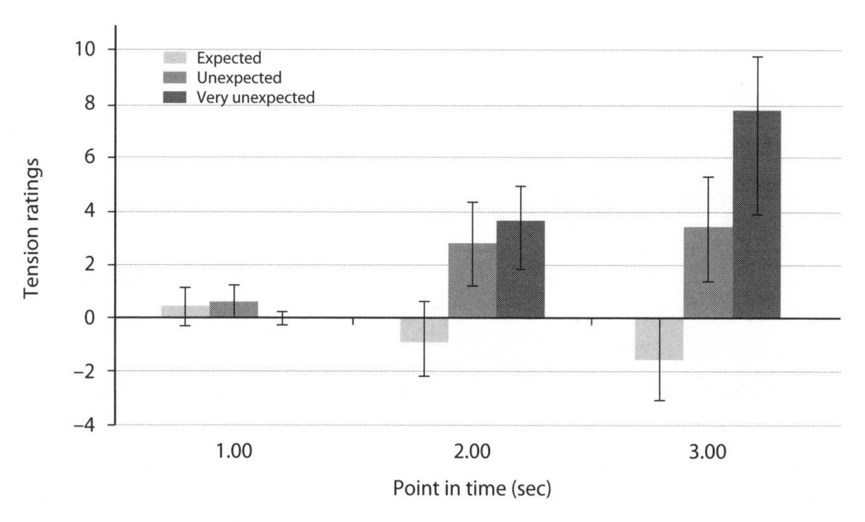

Figure 4.4 Mean and standard error of tension responses to chords of different expectedness over time. Tension ratings significantly increased with harmonic unexpectedness.

Source: Steinbeis et al. 2006.

pression, and resolution. For Dewey, like other drive reductionists, emotions often look a bit like hunger, and behaviors serve to reduce them, eradicate them. What is valuable in Dewey for the present discussion are the diverse ways in which he reveals how bodily sensibility and visceral input figure into adapting to an environment, including aesthetics.[85]

Dewey's view of learning, whether aesthetic or otherwise, is one in which the failure of an expectation initiates the process of learning, a cognitive behaviorism.[86] For example, this view of aesthetics is explicated in Meyer's 1956 volume, *Emotion and Meaning in Music*, in which the discrepancy model of learning figures throughout, as well as in the conflict theory of human appraisals and meaning.[87]

The concept of meaning is itself rather difficult.[88] Certainly meaning in music is not necessarily the same as meaning in language. Thus, a possible way forward is to take the pragmatic route, and simply to refer to a family of related events that are linked to music and for which we talk about meaning. The breakdown of an expectation results in the search for a solution, as seen in musical expectations and their fulfillment.

Expectations and Musical Order

The orientation of an expert, to be sure, is to the syntax or the form of music, and this is usually couched in terms of tonality of musical composition.[89] In some experimental contexts, syntactic forms take precedence over emotional imagery, as is suggested in Meyer's model. I would not suggest that syntax predominates over other forms of information processing that permeate our musical experiences, but for experts, particularly extreme experts (professional musicians), that is a different matter. For example, in experimental studies, musical experts tend to prefer syntactical understanding, compared with novices, in their expectations of form and play (table 4.3).

Several studies have contrasted the fact that non-syntactical experiences of music tend to be devalued by musical experts; on the other hand, high experts tend to enjoy the atypical in musical composition (table 4.4).[90] Even novices, though, are sensitive to syntactic atypicality.[91] The search for something new, original, some form of discrepancy, evokes cognitive pleasure that permeates cephalic expectations. Again, the emphasis on syntactic structure and

Table 4.3

Aesthetic Profile

Aesthetic Dimension	Rating
Syntactic	1.41
Referential	0.06
Sensual	0.47
Emotional	0.76

Source: Smith 1997.

TABLE 4.4

Atypicality: Transformation, Unusualness, Complexity

	Novices	Experts	High Experts
Preference	Harmonic Prototypes	Indifferent	Atypicality

Source: Smith 1997.

music does not mean that other forms of musical experience are not within information processing systems.[92]

Feelings of pleasure are, in many contexts, linked to familiarity. We enjoy the familiar; it is reassuring. Music is a good example. We come back to what is familiar during times of crisis for reassurance. We listen to a song over and over; the familiar reassuring as we embrace something from the past and perhaps also explore and find something new. Our cognitive capabilities and generative creativity allow us to explore and find more to enjoy in the music, to derive pleasure from the familiar and at the same time to find comfort in it as we explore interesting features; or we can simply enjoy the sensory state.

A trade-off between what is familiar and what is novel pervades much of human and, no doubt, animal experience.[93] Exposure tends to evoke enhanced preference for that to which one has been exposed, or with which one is already familiar (e.g., "oldies" music, comfort food, or a known and loved person).

We derive pleasure from music for many reasons—"pleasure in the plural" as Huron has noted. The pleasure we derive from music includes a capacity to prepare our mind to hear sound, amidst conceptual/auditory coherence, endless beauty in audition, and deep social symbolic meanings. We come prepared with biological prepotent cephalic structures to the allure of the auditory, to the musical, to song, and sound rich in meaning, with an allure to complete and fulfill expectations and experience pleasure.[94] Music is inherently familiar to us, even if it contains some unpredictable elements. This could play a role in why it is pleasurable to us. It is like eating variations of your mom's chicken soup . . . maybe a bit surprising, but always comforting and delightful. But this is only part of musical experience.

Feelings of warmth and intimacy can predominate in the familiar. Yet again, one does not have to set this up in terms of cognition versus visceral preference. Preferences are coded in representations in the brain, a highly important form of information processing that serves us well. Desires presuppose information processing about objects that are preferred. Bodily feelings are important sources of information, which is why so many investigators have emphasized them.[95] It is important to gain insight into whether a piece of music elicits chills, or calmness, and how different phenomenological experiences vary with an underlying physiology of the bodily experience.[96]

Perhaps this way of examining events in music is somewhat similar to that of couching cognition and emotions. While cognition is centered squarely within syntactic structure, even those unable to appreciate fine syntactic form may still enjoy the great works in music. Several factors go into musical proclivity; these also figure in composition, including:

- constitutive rules;
- weighted values;
- prominence rules; and
- timing, chord, and tonal rules (see figure 4.5).[97]

Three features stand out: phonology, syntax, and semantics. Cephalic capabilities underlie musical sensibilities amidst a sense of problem solving and detection; the broader theme is the creation process.[98]

Moreover, as Dewey understood, in *Art and Experience*, problem solving is endemic to aesthetics. Music is no different; unconscious inferences, expectations, and the generation of habits underlie the perception and production of musical experiences.

The move away from any perfection holds for the aesthetics in this endeavor. "Satisficing," a word Herbert Simon introduced, applies to the aesthetics of music as well as to the sense that pervades other avenues of human endeavor; the issue is not perfection, but being satisfied given the context and the expectations.[99] One strives not for mythological perfection, but for solutions that are good enough in everyday life, and for "musical heuristics."[100]

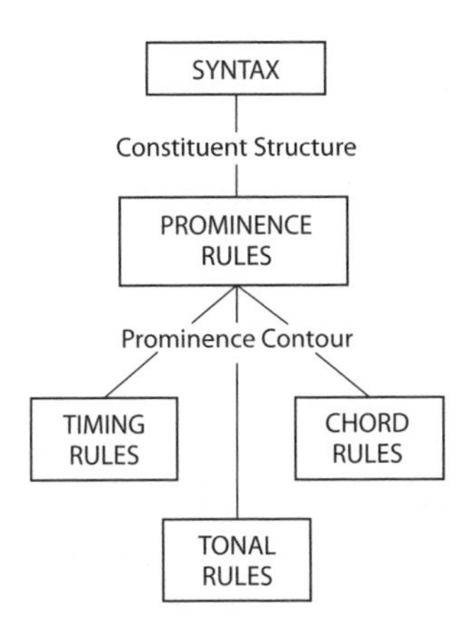

Figure 4.5 Structural features in music.
Source: Sunberg and Lindblom 1976.

As Huron notes, "Like the Pacific bullfrog, experienced listeners of Western music rely on patterns that are serviceable, but not exactly right."[101]

Musical Syntax, Discrepancy, and Brain Activation

Regions of the brain that may underlie musical syntax, probability judgments, and responses to novelty also underlie aesthetic experiences.[102] There is no extra area of the brain that evolved exclusively for aesthetic judgment. Thus, for instance, Broca's area is known for its involvement in the processing of the formal aspects of language, but it is also involved in the syntax that underlies music. There are formal cognitive elements in musical structure;[103] syntactical hierarchical structure dominates both language and music.[104]

Rhythmic sensibility is inherent in language, while melody is codified syntactically in music.[105] Thus while language and music are not the same, they certainly share similar neural origins.[106] This is evidenced by the fact that children with defects in language often show similar problems in musical syntactical competence. Syntactical relationships underlie both language and music.[107]

Interestingly, a variety of brain imaging studies have linked the activation of Broca's area, in addition to other regions including the ventral premotor cortex, to musical syntax.[108]

Premotor and motor regions of the cortex, as well as basal ganglia, are activated in response to music, a phenomenon perhaps more enhanced in experienced musicians with specific expectations.[109]

Some of the regions relevant to syntactical processing are active when a person is listening to rhythms.[110] In one brain activation study (using magnetoencephalography), unexpected or discrepant musical syntactical structure elicited greater activation of Broca's area (and the homologous right side) than a musical composition that was reported as syntactically predictable.[111] This holds for both rhythmic and tonal discrepancies, in addition to semantic discrepancies.[112] This region of the brain is generally responsive to syntactical musical expression. The authors of the above study suggested that the left region (pars triangularis) is more involved in the processing of language, and the right side (pars opercularis) in the processing of musical syntax.

In other related studies, syntactical discrepancy has been linked to event- or activation-related brain potentials for both music and language recognition. The P600 event-related potential was initially linked to language syntax, but it has now been demonstrated that the P600-evoked potential is linked to a broader class of syntactical organization in the brain (e.g., music).[113] The neural mechanisms for the organization of musical judgment are not identical to those of human language expression, but there are interesting overlaps, one of which appears to be the activation of Broca's area.[114] And it is regions like Broca's area that are essential for capturing the syntax that underlies the organization of action, perception, and expectations.

So what are the implications of these findings for our understanding of music? What do they tell us about how we perceive and appreciate music?

Statistical Inference and Music

Clearly, humans possess diverse cognitive mechanisms for appraising probabilistic relations. The history of understanding human reasoning has highlighted this ability, in spite of its imperfection and biases. Human beings use probabilities to assess the likelihood of events, but hardly ever do so perfectly.[115]

Statistics are embedded in practices in which projectability of more entrenched reliability and predictive patterns are depended upon and noticed when they break down.[116] This view underlies reasoning about musical expectations, and to some extent about art more generally, in which a mixture of expectation and discrepancy is held to underlie the aesthetic sensibility.[117] The mental tools used in statistical reasoning also underlie diverse forms of musical competence, replete within a social context of playing with others. They permeate the cognitive processes involved in listening to music, and motivate our response to violations of musical rules.[118]

A critical issue in the logic of statistical reasoning is what constitutes support for a hypothesis. A vast array of ratios and likelihoods is required.[119] Bayes' theorem is one formal tool to scrutinize how frequency information is related to probability judgments in attempting to link relative frequencies and sample size.[120] Bayes' theorem penalizes, among other things, unnecessary model parameters and thus encourages simplicity. We arrive at some ideas rather quickly, a fact that Peirce linked to abduction (genesis of an idea, musical or otherwise) and instinctive responses, rapid and heuristic.[121] There is an experimental sensibility that pervades the creation of music that is constant across cultures and throughout the world.[122]

Keeping track of events is enhanced by the expansion of our sensibilities in the age of information and the cognitive revolution. This expansion is clearly emboldened by the same cataloging and quantification of pockets of information that underlie musical competence and composition.

Musicians come with different musical strengths, and if one does not have a "feel" for the music, one is never going to be a really good musician. Similarly, if one does not have a feel for phenomen as a scientist, one is never going to do good research. Nothing takes the place of this feeling for something. It is certainly basic to the human knower experiencing the world. While expectations linked to probability judgments are only part of the story for musical aesthetic appreciation, they still seem to be elements of the feel for phenomena, which requires a number of information processing systems.[123]

Abduction and Music

Abduction (genesis of an idea) is always constrained by context and ecology, as is musical composition and creativity; problem solving in context pervades musical expression. The creation of a new score, foraging for coherent expression with others in a musical moment, is constrained by cognitive capacity, individual competence, and that moment of pulling things together in the creation of something new amidst the resources available (see figure 4.6).

Our sensory capacity is keen to detect objects that afford sustenance or harm.[124] Fast forms for detecting information reflect diverse heuristics; fast ways to solve problems, both specific and general, developed in the evolution of our brain. We come prepared to associate a number of events linked by causal building blocks in cephalic structures with worldly events. Ecological sensibility and rationality describe readily available heuristics well grounded in successful decision making.[125]

Classical pragmatists understood this to mean that ecological rationality places decision making and the use of statistical features within predilections about numbers and representations of frequencies in real contexts, as we do in music.[126] Moreover, we lean less on memory being strictly in the head by contextual cues to which we expand our capabilities, in musical settings and other forms of human expression.[127]

As the musicologist Eric Clarke discusses in describing Jimi Hendrix's performance of "The Star Spangled Banner," recorded at Woodstock in 1969, there are different attributes of the perceived meaning of music. Sound, structure, and ideology coexist and are simultaneously available in understanding Hendrix's rendition and the infusion of blues, rock, and civil strife in the 1960s, and for the abductive expression of creative expansion from the familiar to less familiar as our horizons expanded. In the same way, numbers exist in the context of contact. Math, like art, underlies all forms of human expression.

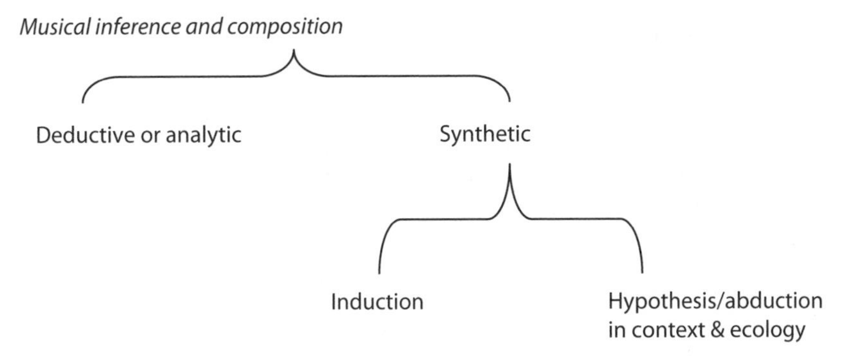

Figure 4.6 Inferences in music.
Source: Adapted from Peirce 1868, 1992.

My colleague Kate Mertes has a memory that is relevant: "My niece used to swim competitively in high school, and they always played the anthem at each meet. One time a 16-year-old bassist played the Hendrix version. The parental viewers were totally buzzed, even the ultra-conservative military parents; we were all sixties babies who remembered the historical totality of the Hendrix version. The kids were all a little shocked: 'Is that really a respectful way to play our national anthem?' Really shows you the importance of context."

Numbers and Music

A sense of numbers is a general feature of our cognitive capability. Numerosity is easily expressed and easily triggered even in the untutored, as Plato demonstrated in his depiction of the slave Meno, never exposed to geometry, who nevertheless demonstrates some capacity for it. We are prepared quite readily to express degrees of mathematical competence; but, only a few are really good mathematicians, whereas most of us are good language users. The native gift of hardware endemic to our cephalic capabilities varies in kind and degree, and of course, the link to music is pervasive.[128]

Reasoning about numbers is expressed early in ontogeny. It is a generative cephalic capability that serves many functions.[129]

One suggestion is that cephalic systems embody a "number sense," meaning that number sensibility is an essential aspect of human cognitive capabilities.[130] Moreover, mathematics is more generally linked to our orientation to action, which is continuous with cognitive exploration and problem solving. This is certainly consistent with Peirce's view that reasoning about numbers shares many features with reasoning about syntax, embedded in music in terms of cephalic function.[131]

But mathematics, like other basic cognitive adaptations, inheres in just about everything we do. Mathematics often seems abstract to people, but it is in fact at the heart of the organization of action and perception. For instance, the geometry of getting through a door opening is a statistical computation readily accomplished in everyday action. As Lakoff and Nunez suggest, "Mathematics is a systematic extension of the mechanism of everyday cognition,"[132] and thus mathematics may be the most general form of application of human cephalic capabilities.[133] It resides in commonplace features and is rooted in objects, the physical world, and music.[134] Keeping time in music is a core feature of rhythm, and it, along with tracking time and the like, relies heavily on mathematical sensibilities.

Mathematics pervades music and everyday forms of reasoning. It is inherent in the way we structure music, from space to time, and to predictions about dangerous or opportunistic events. It is taken for granted because of its pervasiveness. As Lakoff and Nunez propose, mathematics involves:

The embodiment of mind. The detailed nature of our bodies, our brains, and our everyday functioning in the world structures human concepts and human reason. This includes mathematical (or musical) concepts and mathematical (or musical) reasoning.

The cognitive unconscious. Most thought is unconscious—not repressed in the Freudian sense but simply inaccessible to direct conscious introspection. We cannot look directly at our conceptual systems and at our low-level thought processes. This includes most mathematical (or musical) thought.

Metaphorical thought. For the most part, human beings conceptualize abstract concepts in concrete terms, using ideas and modes of reasoning grounded in the sensory-motor system. The mechanism

> by which the abstract is comprehended in terms of the concrete is called *conceptual metaphor*. Mathematical (or musical) thought also makes use of conceptual metaphor, as for instance when we conceptualize numbers as points on a line.[135]

All these functions are certainly applicable to music. So, a sense for numbers underlies music, is cephalically mediated, and is expanded by memory capacity (perhaps with external props).[136] One region of the brain linked to numerosity is the inferior parietal region of the neocortex.[137] This region is tied to a number of behavioral functions, including movement and intention.[138] For example, in one interesting experiment using fMRI, activation of the caudal region of the posterior cortex suggested that spatial representation and tracking events by eye movements are used for numerosity, or simple mental arithmetic.[139]

Probability reasoning reflects, perhaps, a specific mechanism in the cognitive arsenal for understanding the world, and is anchored to ecological and social context.[140] It is a mechanism that evolved to detect danger and to promote reproductive success, and it underlies our appraisals about diverse events, including affective ones. In addition to survival decisions, probability reasoning has been extended to cover novel domains, including aesthetics.

A legitimate question concerns the process by which mechanisms designed to predict predators, prey, food sources, and sexual success might have become linked to aesthetics.[141] Prediction for one set of circumstances extends to new domains in our evolutionary ascent; our cognitive apparatus extends and expands and now can underlie aesthetic judgments because they reflect a response to violations of expectations.

Conclusion

Emotional integration in music is much greater than simply the syntactical part, albeit syntax is a very important part of the cultural evolution of music.[142] The same sort of issue permeates the discussion of the aesthetics of music as it does the emotions; the cognitive, in this case, the syntactical play, takes precedence over everything else. But again, other forms of aesthetics, such as the sensual and the emotional, are all part of the information processing and aesthetic sensibility of music.[143]

With regard to an aesthetic appreciation of music, bodily reason cuts through the distinction between images and syntax, since both are part of the information processing systems that underlie bodily sensibility. "Music is the great link between the sensuous and the intellect,"[144] but the expansion of what we can experience occurs through musical composition. Like the telescope or microscope, the horizons for seeing, hearing, and sensing are expanded; it is a mind in a body. The intellect is already in the sensuous; there are no precognitive givens. But there are diverse codified habits (musical capability) tied to social context and performance in which we need to have less in the head and more in the context of performance.[145] Recognition, as well as storage, are important cognitive adaptations that play a major role in musical sensibility.

Aesthetic appreciation, in its myriad forms, is an essential part of the human condition and is inextricably linked to a sense of well-being.[146] The evocation of broad-based emotions is tied to aesthetics. Human aesthetic pleasure requires information processing and problem solving. The organization or structure of music has long been linked to information processing systems, and these systems in the brain are oriented to discrepancies in musical syntax or form.

The history of aesthetics is rich and wondrous. The diverse expressiveness of humans, both in appetitive and consummatory experiences, our creative and constructive sides, is vastly rich in expression. Aesthetic experiences pervade human life. Demythologizing aesthetics means we have to recognize the pervasiveness of aesthetic sensibility, the cognitive resources that are allocated to it, and the common cognitive and neural systems that underlie its sensibility. Fertility rituals involving music and dance, the rain dance, drums and flutes, and the movement and stories of the earliest human aesthetic expressions are all linked on a trajectory with our modern Broadway musicals, dinner theater, concerts, symphonies, and ecclesiastical rituals across the human condition.[147]

Evolution favored increased accessibility of core cognitive systems, allowing them to be extended in diverse domains of human expression. Aesthetic appraisal gained its adaptive significance, perhaps, by extending from our exploration of novel and discrepant events involving those around us and the world to which we are trying to adapt, to carve out a meaningful existence. Aesthetic judgment exists amidst the information processing systems in the brain that underlie perception and are oriented to novelty, familiarity, and syntax.

Discrepancy (or novelty) detection is a core motivator of brain processing in both musicians and non-musicians.[148] The recognition of musical discrepancy is expressed early on in development. Expectancy effects permeate the organization of the brain and are linked to appraisal systems for gains and losses; the events are embedded in information processing. But discrepancy detectors are only one system; aesthetic sensibility is much broader than simply the detection of discrepancy and the appreciation of filling in the information that is needed. Sensing the syntactical and informational content of the aesthetic object is also important for aesthetic appreciation.

Novelty detectors are operative across the human experience, and in a number of regions of the human brain. Several areas of the brain, including the amygdala, the basal ganglia, and regions of the prefrontal cortex, as well as Broca's area and its analogue in the right hemisphere, are importantly involved in some forms of aesthetic judgment as well as the detection of discrepancies. Moreover, activation of the left prefrontal cortex is linked to music that generates feelings of joy and happiness, while activation of the right prefrontal cortex is linked to music that generates feelings of fear or sadness.[149]

Not surprisingly, in one PET study, a number of other brain regions that are linked to emotional appraisals were activated by music; these include the ventral striatum, the amygdala, and the orbitofrontal and ventromedial prefrontal cortices.[150] These findings are consistent with a large body of evidence in which laterality of frontal cortical expression is linked to positive and negative emotions, and many of these same regions also underlie detection of novelty and discrepancy, and whether to approach or avoid objects.[151]

An interesting question is why, from an evolutionary point of view, we might have aesthetic appreciation. I would suggest the obvious: to mobilize approach and avoidance of objects that we like or loathe.[152] The mechanisms for aesthetic appreciation utilize pre-existing neural and behavioral systems in the organization of behavior.

CHAPTER 5

Musical Expression, Memory, and the Brain

Music is a universal feature of the human condition. All human cultures are rich in song and instrument. Along with this, dance is equally universal. Music is rich in content and expresses a shared sense of the world. The amazing thing remains the vastness of our arsenal of expression, both through music and other forms of communication, in our efforts to create social cohesion and to make sense of the world around us. Music is also one vehicle to enhance memory.

The generative auditory/song processes produce such awe-inspiring or sublime displays, to evoke an eighteenth-century western conception familiar from Coleridge and Kant. The sense of the sublime emerges, as Kant noted, from the experience of beauty.[1] In music this sense of the beautiful can seem unbounded, but it arises from generative cognitive processes rich in teleology and purpose, replete with freedom of expression within constraints amid what Kant called the "free play of the imagination." Cognitive capability is inherent in a brain/body prepared to communicate through song. The grammar of music sets the conditions along with the requisite anatomy.[2]

This chapter explores the issue of musical sensibility as an instinct and continues to address the cognitive and neural capabilities that underlie music expression, including diverse forms of memory.

In particular, working memory is examined as an evolutionary trend that expanded our problem solving and social expression.

Music, Memory, and Instinct: Prepared Capabilities

On the voyage that conceptualized an important idea already circulating in Victorian culture—adaptation and natural selection—Charles Darwin spent quite a bit of time studying the phenomenon of song. He was keen to understand song as a biological feature: "It is probable that the progenitors of man, either the males or females or both sexes before acquiring the power of expressing mutual love in articulate speech, endeavored to charm each other with musical notes and rhythm."[3] Darwin posited that song evolved with communicative capabilities, which extended for some species (e.g., songbirds and humans) with great variation.

Musical sensibility is tied to our social instincts. Darwin noted as early as 1859 that social instincts, including song, are the prelude for much of what governs our social evolution.[4] Climatic variation and stability perhaps fostered group formation as an important evolutionary adaptation—linking up with others, helping them, being helped.

William James expanded the concept of social instincts, noting that "They are not always blind," and that they often organize action.[5] As James also noted, memory matters in social instinct. It expands cognitive capabilities, making them both instinctive and more reflective: "Every instinctive act, in an animal with memory, must be accompanied with foresight to its end."[6]

Memory, of course, pervades music, episodic or specific to individuals or cultures.[7] Both practice and memory are embodied in performance, as well as codified in cephalic structures in which context and the cultural milieu pervade the memorial spaces across time and place. The memories are encapsulated, in part, by sequences tied to performance memory and associative memory, in which we remember things in the context of music.

Music is tied to core instincts. Instincts, for James, as they were for Darwin, are intertwined with problem solving. Such instincts can be narrow or broad, as they underlie the organization of action.[8] Or, as Susanne Langer would suggest, all behavior in the animal is rooted in some form of instinct, including diverse forms of learning, ". . . arising from organic sources as impulse seeking expression in motor action and guided to direct or indirect consummation by acts of perception."[9]

Darwin and the ethologist Niko Tinbergen[10] understood that functions can change over time and be put to novel uses. Musical expression requires a wide range of such functions: respiratory control, fine motor control, and other preadaptive features. This figures into song production, an evolution tied to speech and the diversification of our communicative competence.

Thus, in the context of music, we need to stop thinking of it as one single or narrow adaptation and perhaps no adaptation at all.[11] There is nothing narrow about music any more than there is about language. If language is an instinctual adaptation, core features of it would include how readily we express language, how easily we learn it, how contextual and universal some of its features are.[12] Calling language an instinct, as Steven Pinker does, highlights the ease and pervasiveness and dominance of language. Indeed, music has much in common with language, as many people have pointed out, including the depiction of a vocabulary; both are complex and culturally transmitted.[13]

Pinker suggests that music, unlike language, is not an instinct or a core adaptation. But musical sensibility is an innate predilection expressed in all human cultures, and that is instinctual enough for me. Music is also tied to diverse forms of emotionally laden communicative/memory-related expressions. Like a blood pulse through the atrium, a "musical pulse" pervades cephalic adaptation.[14] Language is a bit like oxygen for our species; our evolutionary ascent is linked to this core feature, a marking distinction, as many have noted over the past four hundred years.[15]

Musical sensibility surely counts as something just as fundamental about us as a species. From a simple adaptation there can emerge such lively expression in any culture. Music is indeed generative, structurally recursive, and knotted to grouping. The generative part, the engine of cognitive processes, is a core cephalic feature that is utilized in diverse contexts.[16]

As much as I like birdsong, birdsong (or whale song) and musical capability are not the same. The debate about how similar birdsong is to human song is not about the songs being identical; they are not. In spite of this, as long as one does not exaggerate the implications of studying song in birds, the mechanisms and resources from which it generates are rooted in similar biological and, in particular, cephalic tissues.

Vocal learning and memory are prior to both speech production and song production.[17] While distinct differences between language and music exist, they are both, quite correctly, viewed as universal. They advance evolving communicative skills that cut across diverse behavioral expressions that facilitate and have transposable proper categorical features. In other words, we come prepared to sing, to encode the events and remember them, to play music, and to appreciate music.

Instinct is one way of characterizing it, although perhaps it is not the best way. After all, no two instincts are the same (thirst, hunger), and hardly any two investigators can agree exactly on what the term even means. In spite of this, Tinbergen gave great descriptions of the diverse forms of adaptive responses that he claimed (and I would agree) are rooted in the concept of instinct. Indeed, Tinbergen, following Darwin in his beautiful book *Instinct*, set the stage, along with other ethologists, for careful behavioral studies grounding music in biology.[18]

Instincts, for William James, are bound to problem solving. Such instincts can be narrow or broad, but they underlie the organization of action. We indeed have a drive to learn, but calling that an instinct can be misleading. The concept of "instinct" is quite loaded in our lexicon. I prefer to say that these are ready capabilities linked to experience, culture, and ability—something we learn quite readily (e.g., learning to associate a taste with visceral distress).[19] Music is one of those things that we do spontaneously, reflecting brain machinery linked to communicative functions, enlarged and diversified across a broad array of human activities.

Perhaps a better metaphor for our diverse capabilities is to refer to a toolbox of cephalic capabilities in which music is one hammer or wrench that evolved as a communicative function.[20] Certainly from the point of view of social communications and memory, music is an important social cohesive tool, laden with binding emotions.

Birdsong, as well, is an inherently communicative expression, richly linked to species specific characteristics.[21] Primates expanded the line from hoots to calls to song; but interestingly, only some primates sing.[22] Vast arrays of species signal others through sound. It is quite commonplace in the animal world to find that specific calls are linked to different kinds of objects. Predators and danger are linked to one kind of call, as opposed to signaling about non-threatening objects or perhaps even desirable objects. These patterns were demonstrated in playback experiments when the sound was reproduced and behavioral responses were observed. Some monkeys, for instance, have a specific call for eagles overhead and a separate call for snakes on the ground below. Calling and hollering, however, is commonplace across primates and, like song, these are ultimately all social in function, rich in cognitive resources and memory functions.

Cognitive Capacities

The cognitive revolution not only overthrew the narrow view of what terms to use in describing behavioral adaptation, but also tied cognitive systems to visceral adaptation. It is truly a revolution in the history of biology. Darwin and others would be dumbfounded at the concept of the separation of cognitive systems from action, from visceral peripheral intimacy in the orchestration of adaptive response. The origins of song are embedded within such systems.

Music is inherently tied to movement and time (see chapter 7). It is also closely linked with cognitive events, adaptation, sensory experiences, and emotional sensations.[23] Bodily sensibility underlies musical sensibility; music transmits affective information to us, and the resulting bodily sensations give us information about the message of music.

The general depiction of the brain is focused on circuits and their organization in functional context. Song, like speech, is interactive both in terms of its social uses and its recruitment of various physical and cognitive resources. In music, there is no separation of mind and body; it is rather a great expression of the connection between the two.

Music warms and cools, tranquilizes and invigorates. It soars and descends within time, expanding our sense of time as we merge, our minds engaged with no Cartesian separation. The social context highlights what Dewey referred to as "lived experience."[24] Peirce, an early proponent of statistical literacy, was also a proponent of semiotics. Cognitive/motor skills are engaged in listening and certainly performing, in watching and simulating, in imagining and transcending.

Meyer indeed held a kinesthetic/expectation conception of musical sensibility, suggesting that neural systems are linked to perception, or a gestalt, or rich expectation in which gaps in information are filled in a context of musical expression. This is a core pragmatist position in which the social/musical self is more broadly tied to anticipatory responses and instrumental capabilities, among other things.[25] The first factor is the fundamental link to others, the second is language capabilities, and the third is the unending creation of instruments for seeing, hearing, learning, all of which are knotted to music. Is music, therefore, a separate faculty?

I prefer to see music as an expression of a set of capabilities, some of which are highly constrained and some not. There is a substrate that makes language possible, involving combinational capabilities, cultural expression, and neural ties. There are specific regions of the brain that underlie language, but these regions (e.g., Broca's area and regions of the basal ganglia) are also involved in a wide array of other events, including music.[26] Using fMRI to measure brain activation, we can see that rhythmic beats activate diverse regions of the brain, including the basal ganglia and the pre-motor and motor regions of the neocortex. The neural and cognitive resources stretch across domains. The resources that produce music are quite distributed. In spite of the specific systems that play key roles in musical production and understanding, song stretches across cognitive and motor systems.

The evolution of science is intimately linked to invention, expanding cognitive resources to ever widening domains;[27] experiment and experience are the pulse of science. In science, we find that through invention and discovery our expression is expanded; the same holds for music. There is no separation, just diverse types of expression. Cognitive motor systems underlie the building of instruments for playing, discovering, and exploring.[28]

After all, instruments like telescopes and microscopes allow us to extend our ability to see and hear; as do many, if not all, scientific and cultural inventions. The domain in which we can explore is advanced. This is also true for musical instruments, in that they expand our domains and reflect the larger cultural milieu in what is available, usable, extendable, and part of the lifeblood of individuals and cultures.

As Dewey noted, the same forms of reasoning pervade all of human expression.[29] Aesthetics is embedded in science, and science in art. He says that a sense of "form is arrived at whenever a stable, even though moving, equilibrium is reached."[30] This experience with the development of instruments for expression pervades musical sensibility, as it does many avenues of human expression. Dewey placed emphasis on the importance of adaptive tools as part of our experience in the broad array of human activity, including music and art more generally.

The building of things, musical and otherwise, pushes us, luring us forward to expand our horizons, a key part of the human spirit of inquiry. These are features of music, but also of all of human experience.

Within this exploration, social connectivity (calling out to others, reaching others, seducing, intimidating, or contacting others) coevolved within the larger cultural milieu. Music is at the heart of this evolution, and every feature of us is within the expression of music.

Music pervades memory systems, and we now turn to the diverse forms of memory and continuity, or in James's language, "The stream of our thought is like a river, . . . and effortless attention is the rule."[31]

Meyer understood the effortless continuity pervading musical sensibility that aids memory.[32] He notes that, "Chopin's prelude Op. 28, No 2 presents a clear example of the establishment of a process, its continuation, a disturbance, and finally the re-establishment of a variation of the original process."[33]

Memory and Music

There are many types of memory.[34] Autobiographical memory is one form that refers to our personal experiences, the semiotic networks associated with our past and references to the vast array of meaning associated with ourselves. The medial prefrontal cortex is only one key region, though a rather large one, associated with self reference, autobiography, and musical sensibility. Indeed, one study found that both music and autobiography are deeply ingrained in the architecture of the medial prefrontal cortex.[35] This may reflect how effortless it is to recall, through music, so many diverse experiences we have had.

Not surprisingly, since emotions are bound to core memory, musical cognition and cognitive motor planning memory are inherent in musical sensibility.[36] Depicted in table 5.1 are some features of memory (procedural, semantic, working, and autobiographical memory).[37]

Intentional action is often taken in the context of actual others, potential others, and imagined others (deceiving others, helping others). Even the most reclusive of us imagines others or lives with others, at least to the extent that we are, perhaps, still intentional and have a background of socialized ambiance—rejected or not. Moreover, the worlds we inhabit and adapt to already contain well-worked practices that pervade the world we embody.[38] The same goes for music.

To relieve the brain from having to store too much information, we derive information directly from the world in which we live. Gibson[39] emphasized action patterns that are released by properties of the world in which we participate; music is one of these features.[40] We are guided by direct perception in tracking music—understanding the trajectory of the cadence, pitch, and general structure. Our perception is richly laden with the appraisal of affective content.

TABLE 5.1

Partial Memory Taxonomy

System	Other Terms	Subsystems	Retrieval
Procedural (musical)	Nondeclarative	—Motor skills —Cognitive skills	Implicit
Semantic (musical)	—Generic —Factual —Knowledge	—Spatial —Relational	Implicit
Primary (musical)	Working	—Visual —Auditory	Explicit
Episodic (musical)	—Personal —Autobiographical —Event		Explicit

Source: Adapted from Schacter and Tulving 1994.

To simplify our cognitive abilities, we learn to tap into the meanings and stability of natural kinds.[41] Therefore, a lot of information processing isn't always necessary. In this sense, meaning comes in the reception itself.[42] Gibson is vague, but his point from the biological side is that the world is already coherent and has inherent regularity or clusters of regular patterns that provide coherence, for which the brain is prepared to receive information. This holds for musical expression.[43] We still need to code memories and meaning; music is not strictly in the head.

Memory is inherent in our experiences of music. Listening to Mahler's Fifth symphony is bound for me to a memory of walking through cold and rainy London while somewhat distraught over a young love, on the order of Kierkegaard's lament with regard to his lost love. While the opera might emphasize drama, we all have our own variations of drama, of which music is integral to the experience and to the memories. The "dream-like" quality of Mahler's music is easily tied to memory.[44]

One vital feature for our evolution is short-term capability. We do much in the course of the day. The social milieu is quite large for most of us, and this sort of capability is linked to a capacity to remember diverse events. Some are autobiographical and some are particular, some are linked to others, and some are specific to certain functions.

As James noted in his great book on psychology written over a century ago, there are diverse forms of memory, short- and long-term, working and autobiographical, semantic and episodic.[45] Working memory is only one memory system that is operative in musical sensibility; one neuroanatomic part of the memory system.[46]

Indeed, like the effect that pre-exposure sentences have on recognition tests, musical pre-exposure has many of the same effects on sentences in memory.[47] Telling a story in music and sentences is captured by diverse cephalic structures.[48] Regions such as the hippocampus, the amygdala, and the anterior cingulate cortex, all classical limbic regions, are activated by music and tied to memory.[49]

Indeed, regions long linked to memory (and also seizures), such as the amygdala and more generally the temporal lobe, have also been noted by clinicians to be linked to "musical auras": a strange familiar feeling that emerges out of a piece of music.[50] Oliver Sacks quoted one of his patients as saying, "The music exploded in my head."[51] These musical seizures are unexpected and "arrive out of nowhere," just like the writing of music.

Sacks notes two things about seizures and music: "I have seen a number of patients with seizures induced by music and others who have musical auras associated with music with seizures—and occasionally both."[52] One patient of Sacks' had seizures that were triggered by a range of music, but mostly familiar music and music tied to memory. Sacks quotes the patient saying that Frank Sinatra "touches a chord in me."[53] What is familiar reflects one's specific biographical experience and then the larger cultural memory in which one participates.

Memory spreads across all human action. Short-term memory may have conferred selective capabilities, including that of musical sensibility, as one way to enhance the recovery of important items and events.[54] One essential neurotransmitter linked to memory is norepinephrine, which is important in consolidation from short-to long-term memory.[55] Short-term memory is shrouded within attentional mechanisms and broader based informational systems, all of which figure in musical memory—for example, the learning of a new piece, the transition from short- to long-term memory.[56]

Diverse regions of the brain are biased for competition for expression. Indeed, neurons are competing for expression in regions devoted for instance to visual attention; some neurons are suppressed and some are not.[57] Expectations figure in attention to memory. Musical training enhances cephalic capability and expectations that reverebrate in the brain to simple and complex auditory stimulation.[58]

James noted that attention spreads across the brain in a context, with one feature being that "expectant attention, even when not very strongly focalized will prepare the motor centers and shorten that which a stimulus has to perform on them in order to produce a given effect when it comes."[59] Diverse molecules (e.g., catecholamines) accentuate the attentional mechanism, such as attending to meter, and the facilitation of working memory (see figure 5.1).[60]

Cognitive/Motor Systems: Objects, Movement, and Music

At the beginning of this book, I emphasized that both the cognitive revolution and the emphasis on biological adaptation have fostered the view that there are a diverse array of cognitive mechanisms in learning and in memory.[61] Semantic processing reflects different cognitive and neural mechanisms.[62] More than one neural system appears to underlie semantics in the nervous system.

Moreover, from an anatomical point of view, representations of object knowledge, like most functions in the brain, are not simply localized in one part of the brain, but are distributed across the neural axis.[63] However, the representations are not random; they reflect the underlying organization of the nervous system as well as its evolution and function. One important insight into the brain is that semantic processing reflects regions of the brain linked to action and perception.[64] In terms of music, this means that we

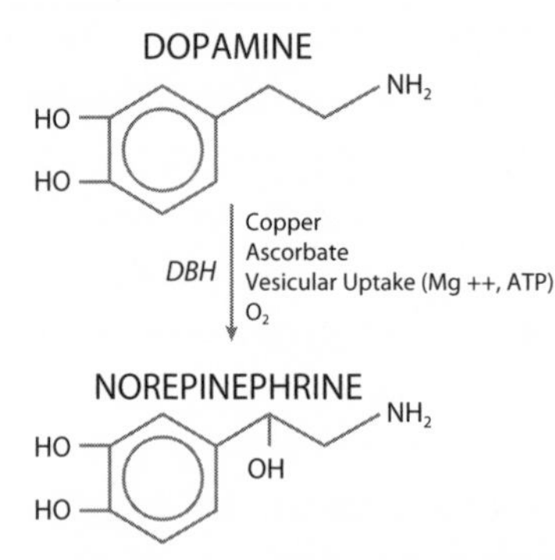

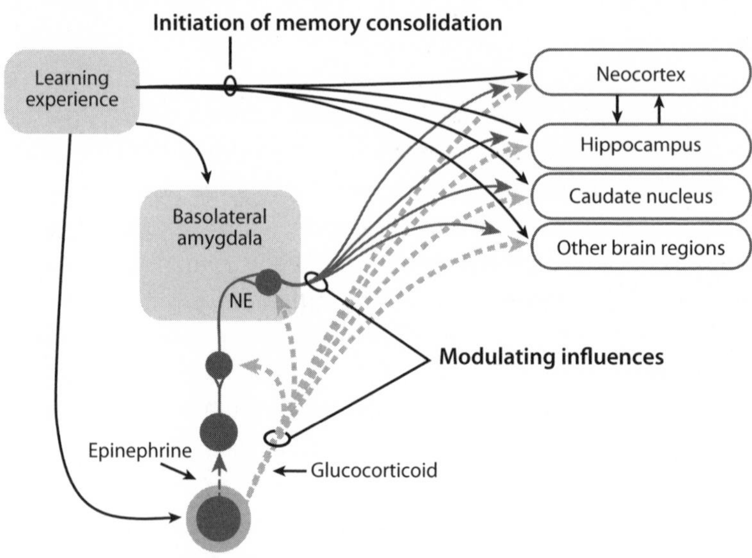

Figure 5.1 Working memory activation and associated neurotransmitters.
Source: McGaugh 2003.

predict and mirror what others do in the context of making music together, directed or not.

Interestingly, in brain imaging studies, action words generated greater activity in the left middle temporal gyrus.[65] This region of the brain is just anterior to a region linked to the perception of motion.[66] Motion is also tied to musical sensibility; music is always going somewhere, constantly in motion.[67] Thus, action is anticipatory, and anticipation underlies the cognitive/motor capabilities of musical expression.

Memory and motor control pervade, for instance, the anticipatory hand position in clarinet players.[68] As an ex-clarinet player who grew up playing Benny Goodman and Artie Shaw, I know well that memory pervades all musical sensibility and expression. The motor memory involved in playing a clarinet is rife with attention and motor control, utilizing the diverse features of cephalic capabilities.

The same holds for all human action and tool use. Without memory there is little action and no self. A cardinal human feature, tool use is tied to communicative gestures, and music is one very important extension of this same expression. Musical instruments are a kind of tool after all, driven by endless integrative streams of sensory/motor expression cognitively linked to sound, context, social interaction, and memory: the memory that is required to play the songs. Of course, the instruments, as tools, are extensions of ourselves, and of our bodies.

Researchers have asked subjects to identify drawings of animals and tools. These classes of objects were chosen because the distinctions we make among four-legged animals usually rest on their physical differences (size, color, and the like), whereas our distinctions among tools are based on their functions. Naming animals (but not tools) activated the medial aspect of the occipital lobe. The activation of the occipital cortex reflects a reactivation of primary visual areas, which may arise from the need to identify an object using the relatively subtle distinctions of physical features.

Naming tools (but not animals) activated the left middle temporal gyrus, the same region that is activated in generating action words associated with objects.[69] The authors suggest that this area may organize stored knowledge of visual motions and their uses. The authors conclude that identifying tools may be partly mediated by the areas of the brain that mediate knowledge of object motion and use, and are close to sites that are active when we perceive motion and use objects. Music is bound to motion and movement (e.g., the clarinet). These same regions of the brain are activated by music.[70]

Motion is a fundamental feature of musical sensibility, the metaphorical expansion that underlies the organization of action and perception. Music, language, and motion are inextricably intertwined and expand our experiences.[71] Using simple geometric shapes to depict possible social contexts elicits greater activation (measured by fMRI) of the anterior region of the temporal lobe, the amygdala, and the ventral medial prefrontal cortex, the same regions linked to motion. These regions are critical for the detection of agency and self-generated movement.

The sense of motion and movement is a fundamental feature in most forms of musical experience, within which emotion is tied to movement.[72] The appraisal of movement and tempo are expectations of the music, locked together into coherence, and are a fundamental feature in the coherence of perception and action. A feature of these events is the so-called mirroring effects of action and perception[73] that underlie social meaning and the understanding of the intentions of others in the playing of music and for discerning the intentions of the composers: "thought is a thread of melody running through the succession of our sensations."[74]

The "mirror neurons" (sets of neurons that fire when a person performs an action, imagines an action, or watches another person do the action) that are tied to musical experiences indicate that viewing and doing are a joint exercise.[75] This is importantly involved, no doubt, in the social cohesion in playing music together, singing with each other, or playing in a band with one another. Mirroring action would be vital in the context of musical expression and sensibility. Of course, mirror neurons are not specific to musical expression, just like dopamine and the other neural systems that have been highlighted. Nonetheless, they play a key part in musical expression.

In music, perception and action are joined.[76] Think, for instance, of the planning and action sequences in musical production.[77] Again, that does not mean they are not separable.[78] Obviously, they are and importantly so, but these sorts of anatomical findings point to common underlying neural mechanisms. Diverse premotor and motor regions are activated in thinking about and playing music; networks for integration of auditory and other sensory systems are the cephalic fabric.[79] The transduction of "sound to significance" is also cross-modal.[80]

We are such visual animals that we can see or imagine seeing even if we become blind.[81] In the same way, I can imagine a piece of music before playing it; and in doing so, I activate many of the same neural regions that I would if I were playing. The important evolutionary and functional point about the brain is that viewing the features of an object activates both visual and motor regions. One region reflects sight, the other action or movement. Both can be active whether we are looking at movement, thinking about movement, looking at tools, or actually moving ourselves through space. We appear to have a predilection to discern motion. Action categories (e.g., animate or inanimate) and functional relationships (use of an object) may figure vitally in discerning motion and causation, and certainly the direction and movement that permeates music.[82]

Musical sensibility has agents (musicians) who are quite clearly animate. We can sense in music the intentions of others, the motion associated with agency, and the fact that others permeate the musical world in which we participate and listen as we are "moved" by the music. Motion is vital to both imagining an action and the action itself.[83]

The anatomical/functional relations have relevance to considerations about regions of the brain and social knowledge. It may be the case that the same brain regions are activated when a person acts or imagines acting in an intentional way as when that person is attributing intentional action to another person. It may also be the case that these brain regions contribute to the organization of action.

Neural and cognitive processing are performed outside of our awareness. What may be learned at first, such as meter and rhythm when playing the piano, eventually becomes habitual, resulting in the acquisition of musical cephalic literacy.[84] Then the focus shifts; the neural order, finite in scope, calls into attention what it needs to attend to in the musical experience, to the musical semantics. We do not have privileged access to the mechanisms that underlie semantic processing in the brain. Semantics are inherently social interactions, however. Even in the solo of the individual performer singing or playing, we are always calling out to others when making music.

Not surprisingly, there are shared and distinct regions of the neocortex (e.g., Brodmann's area, precentral gyrus, inferior frontal gyrus) linked to singing and speaking with some common transfer

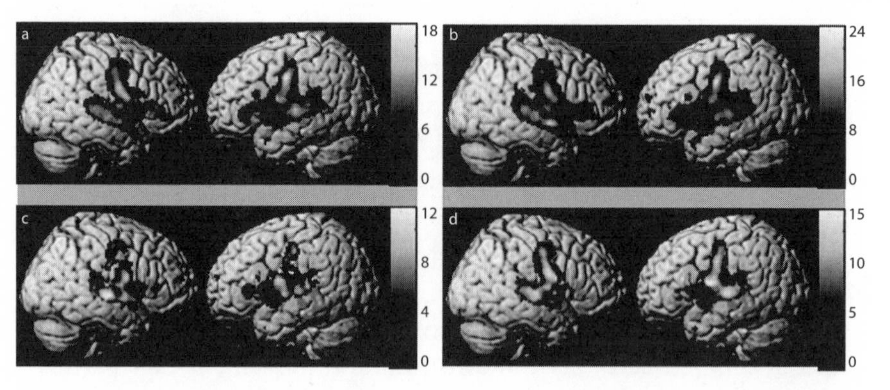

Figure 5.2 Shared regions of the brain that underlie singing and speaking: (a) brain activation while speaking; (b) brain activation while singing; (c) brain activation while humming, each contrasted against brain activation during silence (d).

Source: Reprinted from NeuroImage 33:2. Õzdemir E. Norton A., and Schlaug G. "Shared and distinct neural correlates of singing and speaking." pp. 628–35. Copyright (2006) with permission from Elsevier.

effects from training on one to the other (figure 5.2).[85] The regions that predominantly overlap are the pre- and post-central gyrus, the superior temporal gyrus, and the superior temporal sulcus.

Musical Experience and Changing the Brain

Evidence suggests that the brains of musicians and non-musicians are different. Music shapes the cephalic encoding of information processing across different levels of the brain, from brain stem to cortex.[86] Indeed, early musical training affects children's linguistic expression, and perhaps they are more sensitive in neonatal development and on multisensory functioning.[87] Moreover, musical training enhances auditory capability more generally by impacting cortical and subcortical regions.[88]

In one study, for instance, grey matter differed between the two groups in the motor, visual, and auditory cortex.[89] This may be due to enhanced neural connectivity. One set of studies suggests that in the corpus callosum, the main commissures between the two cortical hemispheres are greater in musicians versus non-musicians.[90]

In addition, intra-temporal lobe connectivity is increased in musicians with absolute pitch.[91] That means that hearing tones more acutely is associated with greater inter-temporal neural connectivity. Based on this information, it would appear that several regions

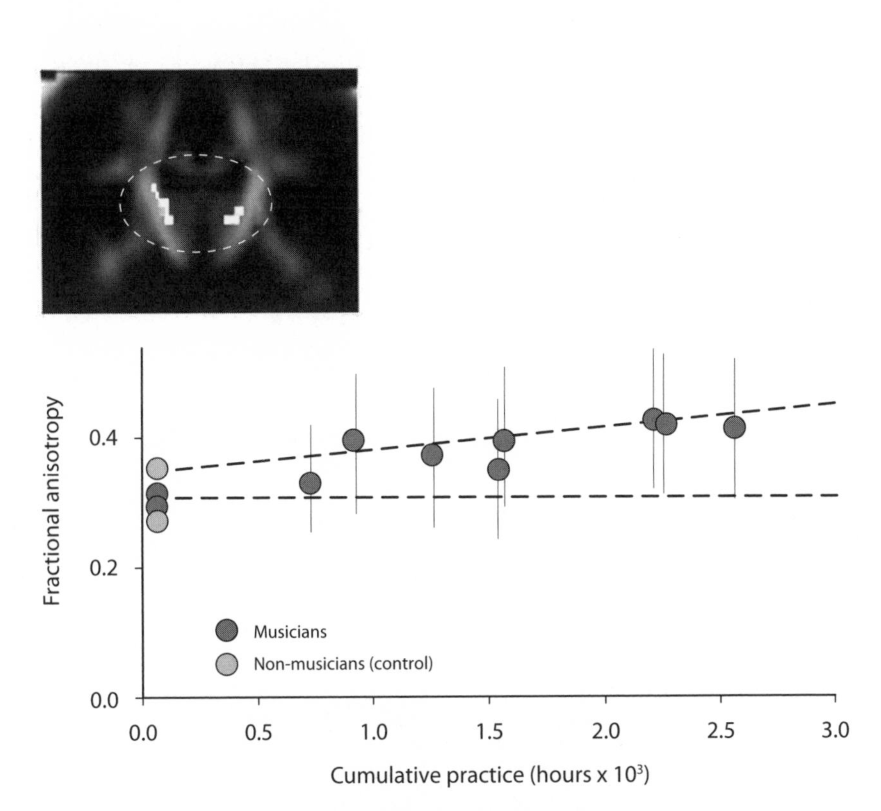

Figure 5.3 Increased white matter in the internal capsules of musicians, represented by functional anisotropy as demonstrated using diffusion tensor imaging, which is represented by the circled region in the top image. These findings indicate that musicians' brains differ structurally from non-musicians' brains, which may be related to the onset and/or amount of musical training.

Source: Reprinted by permission from Macmillan Publishers Ltd: Nature Neuroscience, vol. 8:9, Bengtsson et al., "Extensive piano practicing has regionally specific effects on white matter development." copyright 2005.

of the brain are altered and expanded by the hours of musical practice typically exercised by musicians. In fact, the actual extent of regular musical rehearsal practice is positively correlated to the degree of neural connectivity (figure 5.3).

The auditory cortex and the auditory systems more generally are intimately tied to music and hearing, including speech and song.[92] Damage to these regions disrupts song capability.

Music is richly organized into lexical networks of musical meaning.[93] One suggestion is that the left hemisphere, especially the superior region and surface of the temporal lobe (Heschl's gyrus), is tied to speech, and the right side is tied more to tone (figure 5.4).[94] In two studies, for instance, the grey matter in the right cortical area was significantly greater in musicians than in non-musicians in several parts, including the precentral gyrus and the superior parietal cortex.[95]

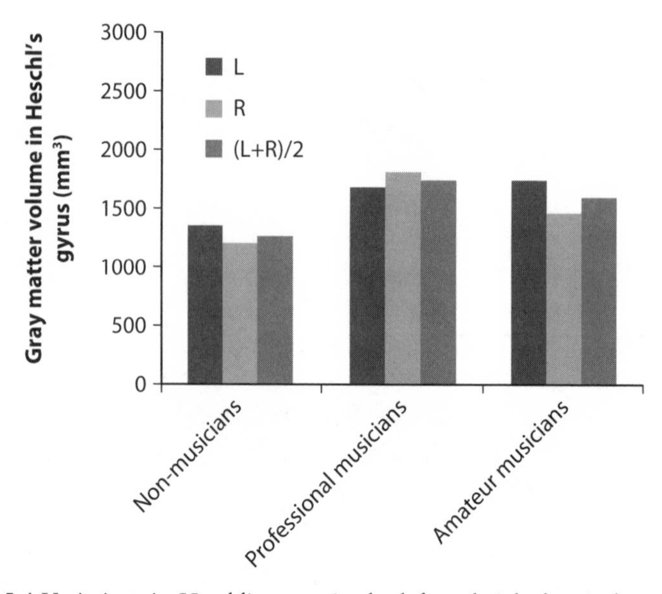

Figure 5.4 Variations in Heschl's gyrus in the left and right hemispheres across three different groups.

Source: Schneider et al. 2002.

Premotor Regions and Musical Expression

The premotor regions and the anticipatory cephalic organization of human action are linked throughout to musical expression. Neural action between premotor regions, auditory systems, and motor output are pervasive in musical expression and the organization of action (see figure 5.5).[96] The dorsal premotor region in particular is knotted to metrical musical sensibilities.[97]

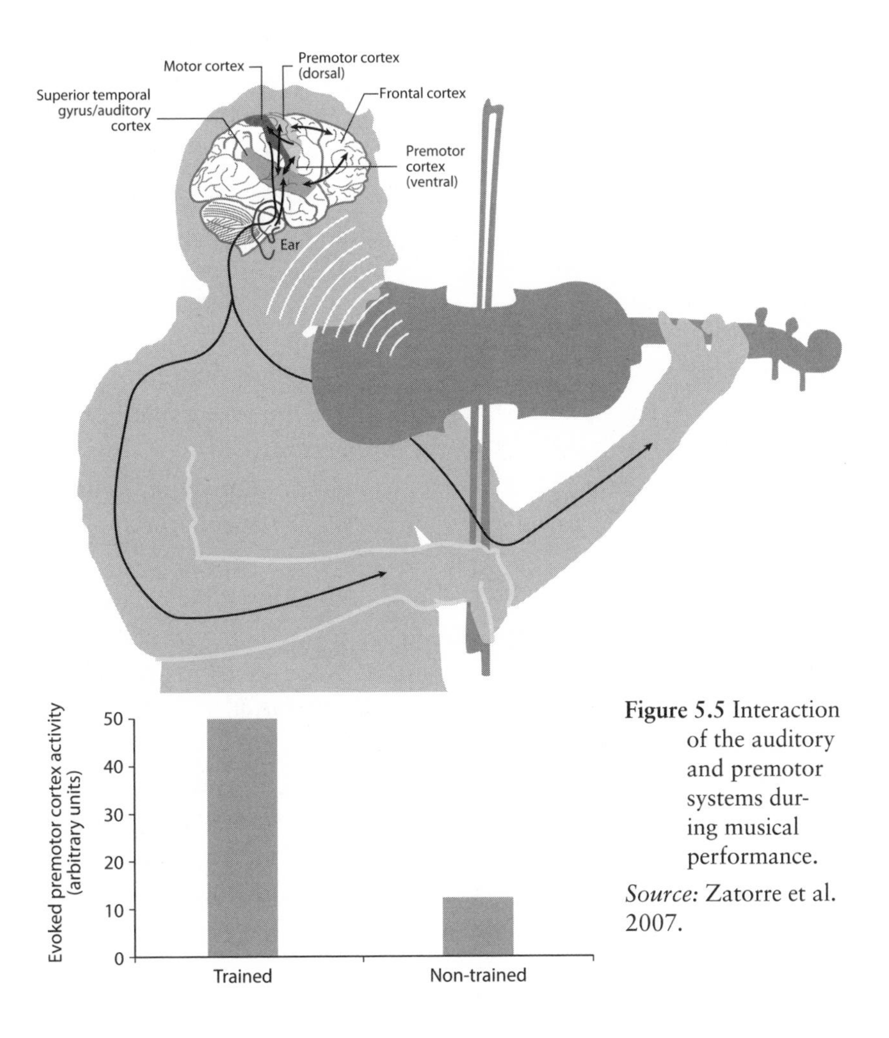

Figure 5.5 Interaction of the auditory and premotor systems during musical performance.

Source: Zatorre et al. 2007.

Cognition is pervasive, underling all forms of musical sensibility, in addition to affective appraisals.[98] They range across a wide array of brain regions, including that of premotor regions of the neocortex.[99]

Conclusion

Memory is obviously not unique to music, yet it is an important aspect in our experience of music. One feature that is pervasive is that musical memory is grounded in the familiar and, when measured for brain activation, diverse regions are more activated by less familiar music.[100] Expectancy, however, is embedded in both short- and long-term memory, pervading music, and is expressed in children as well as adults with regard to musical memory. The music itself and the memories of it are affectively laden.[101] The melodic and rhythmic accents that underlie the structure of the music, in which variation is ever-present, underlie neural capacity.[102]

Memory, premotor, and motor integration figure importantly into musical sensibility, as they do in all human expression. Diverse forms of psychobiological information processing inhere in music, attention, memory, and the prediction of events. Indeed, all cultures have music, and, therefore, core adaptive capabilities are evident, as are changes in brain function that underlie musical expression.[103]

Musical memories are key markers, a fundamental way in which we remember meaningful events—hence my association of Mahler with a youthful failed love in London. Now, as in many others, this memory holds the sweet sense of love and loss. The memory is built up out of music and place.

Memory pervades human experience and constitutes what we mean by the self; memory is not one "thing," however, just as self is surely not a thing.[104] Memory is connected to our expectation of events, and one adaptation is to externalize memory so that it is not strictly "in the head." Markers in the external world remind us of events and activate recall. Music can be one of those external markers that prod us toward recollection.

Did music precede language, or is emotional speech the origin of music?[105] We do not know, and perhaps we never will, but emotional speech through song is certainly a core human tool. The issue is not to equate music to thought or language, but to see it as an extraordinary form of human expression, playing diverse functions in facilitating social contact and social life.

Music is not a faculty, but it is linked to varied cephalic capabilities, in addition to something that Kant called the "free play of the imagination," which for us is a social capability that allows us to link up easily with others in song, dance, and rhythm.[106] Not surprisingly, there are shared and distinct neural correlates of singing and speaking.

Music is social and oriented to others, to instruments, and to expression. An emphasis is on the body's appropriation of objects and their use, which is vital to music.[107] The body is a vehicle of knowledge, replete with cognitive structure for knowing what is around and what to attend to, learn from, and respond to—these events underlie all of musical sensibility. This sense of body knowledge is well represented in the brain, and is part of the organization of intelligent action, of which musical sensibility is a prime example.

CHAPTER 6

Development, Music, and Social Contact

A sense of music begins very early in infancy. In fact, the discrimination of pitch and other perceptual capabilities are expressed within the first year of life, events believed to be fundamentally linked to social capabilities. It is the social world, gaining a foothold in the life of others, which makes this knowledge essential. Rhythmic engagement also begins in infancy, generating movement.[1] This musical expression is linked to affective needs and diverse forms of social contact.[2]

Affective appraisal is inherent in the organization of action. Some events are more affectively salient than others, and the same is true of music; some forms of music are more divorced from affective salience. Music, like all affective expressions, is laden with appraisals. There is no separation between cognition and motor movement; cognition inheres in movement. But the relationship is different on the other side: not all cognition requires actual movement, although it may require some sense of imagined movement.

In this chapter, we begin with a consideration of normal neonatal orientation to sounds before moving on to developmental disorders that affect musical sensibility, including a form of hypersocial expression coupled with a liking for music, namely, Williams syndrome. We will then return to a more general theme of the book, the evolution of social behavior that underlies musical expression. Finally, we will consider epigenetic and lifelong learning changes in relation to music.

Musical Trajectories

Infants are oriented to melodies and their contours.[3] The world of song provided by mothers and fathers facilitates attachment and orienting behaviors. All human cultures have their sweet lullabies, a very general expression of the human condition rooted in social attachment and social comfort that orients children to musical structure and intervals.

Infants come prepared to recognize objects, both animate and inanimate, in the directory of events.[4] Appraisals run through neural function, affectively rich in limbic activation, in the context of music.[5] So, babies readily recognize the significance in the intonations of the voice, movements, and rhythm, for instance in lullabies. While listening for these cues, diverse regions of the brain are prepared for response-social contact, comfort, and alarm. Indeed, probably a large corpus of affectively laden informative events is connected to these events.[6]

For the neonate, rhythmic movements are tightly tied to auditory systems and to musical sensibilities, of which animate objects and social coordination are primary modes of interaction. Indeed, the orientation of the neonate is toward consonance, not dissonance or atonal music. Thus, the attempts to generate a preference for atonality in the West appear to be an acquired taste. Orientation to dissonance is deeply linked to expertise in music but is not something embedded in the neonate. However, we can learn to appreciate dissonance or non/harmonic music. Therefore symbolic and perhaps non-acoustical features come to predominate, that reflect cognitive play within dissonance in the enjoyment of atonal music. However, musical preference is not strictly bound to the perception of notes and tones, but is a cognitive exercise that can be linked to memories, attitudes, emotions, and cultural/contexual external factors.[7]

The demands of our long postnatal period, during which sets of core cognitive capabilities are developed and expressed, are essential for pedagogy.[8] These cognitive capabilities anchor us to our social milieu, orienting us to others as well as to our ecological and social surroundings in a broad array of musical contexts. The long postnatal period ensures this with inherent auditory/motor

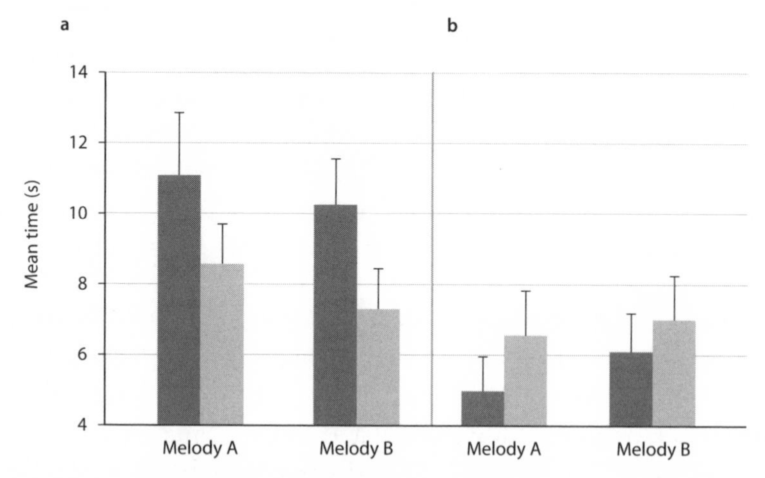

Figure 6.1 (a) Attentional responses are greater in neonates to consonant ■ melodies versus dissonant ■ melodies. (b) Fewer motor actions to consonant melodies versus dissonant melodies.

Source: Zentner and Kagan 1996.

richness and enculturation, training, and entrainment essential for musical learning.[9]

Early training can enhance perception of musical sensibility.[10] An intuitive sensibility for music is just another way to express the fact that the musical systems are core, easily expressed and effortless, although culturally varied, requiring formal and informal training as part of the socialization process amidst the perception of consonance (figure 6.1).[11] Temperament is also a feature in all factors of the human condition, including that of music, as is the lateralization of neural expression with regard to emotional appraisals of music.[12]

The brain is prepared to respond to music (table 6.1). Infants are indeed, in some contexts, able to tune into rhythm more easily than adults, as specialized development in the brain for musical processing begins in the first few days of the neonate's life.[13] Diverse common processing is expressed in musical sequencing.[14] A musical/auditory syntactical template underlies the cephalic capability in the child to engage music early and effortlessly. What varies is the cultural milieu and the individual talent. Like the heart, music is a core feature of humans. Song is a universal trait tied to diverse emotions in which classification schemata and measurement are paramount.[15]

An understanding of musical experiences demands different levels of analysis including genetic, neural, cognitive/behavioral, and cultural. Interestingly, amusics (individuals that are unable to carry a tune) and their first cousins often have the same impairments in the perception of pitch.[16]

Musical sensibility can be spared in developmental disorders, although not always.[17] Learning disabilities have been noted for music itself. Indeed, forms of devolution of function decrease the competence to integrate music into motor expression with tone and melody.[18] This occurs because pervasive links between diverse appraisal systems permeate music, not because one is cognitive and the other is not.

Maurice Ravel, the composer of *Bolero*, is reported to have developed aphasia after a blow to the head.[19] While he apparently felt he could hear his music in his mind, he was unable to write it down. However, other conductors who have developed aphasia and even motor apraxia have had their musical ability spared.[20] Extreme musical devolution such as Ravel's is rare.[21]

Some individuals, however, appear to be amusical; disorders of musical perception are clearly demonstrable.[22] One example is

TABLE 6.1

Early Exposure to Musical Developing and Training

Passive experience	Active experience	Early developing universal constraints
Enculturation to a musical system	*Formal music training*	*Consonance*
Scales	Performance	Temporal regularity
Key	Music reading	Multisensory interactions
Tonality	Explicit knowledge	
Meter	Attention and executive function	

Source: Hannon and Trainor 2007.

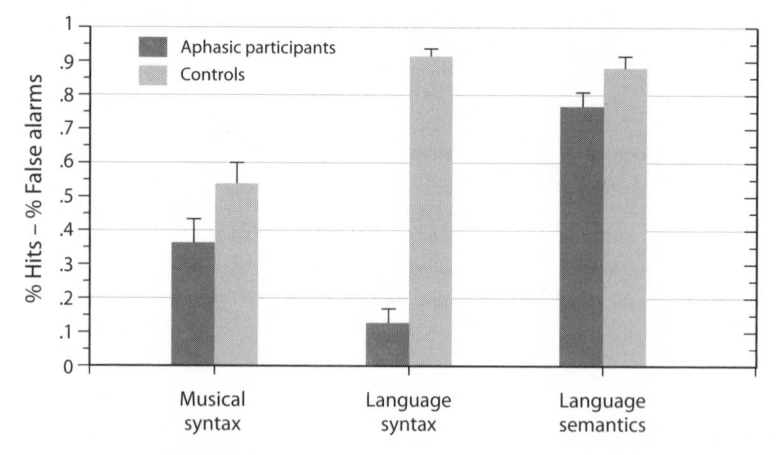

Figure 6.2 Comparison of individual performance on musical syntax, language syntax, and semantics tasks in aphasic individuals and normal controls.
Source: Patel et al 2008.

impaired detection of pitch, or tone-deafness.[23] We do not know much about the possible genes that underlie amusicality, but we do know that diverse levels of analysis underlie any explanation of those with ability for music as well as those without any talent.

Patients with grammatical problems or impaired syntactical competence, along with the fact that the same cortical impairment may not be associated with amusicality, suggests that subcortical regions of the brain such as the basal ganglia may play an important role in musical syntax in addition to more general rhythm (figure 6.2).[24]

Hypersocial and Hyposocial Individuals

Individuals with Williams syndrome share a common genomic marker and other common features. Their general IQ is usually much lower than the general population, and they have great difficulty with numbers and math. Their spatial capability is quite poor, although their linguistic capability is often good.[25] Interestingly, motion processing in individuals with Williams syndrome is not perfect but remains fairly good, suggesting that the ventral stream linked to motion and agency is operative.[26] But the hypersociality associated with Williams syndrome is its most marked feature.

Often described as having "cocktail party" personalities, individuals with Williams syndrome are exceedingly cheerful, associate easily with strangers, and hyperfocus on eye contact when socially engaged. Thus, while deficient in some mental capabilities, Williams syndrome individuals nevertheless have intact and highly evolved human expression, including a strong love of music and a much greater than average expression of perfect pitch.

Autism, by contrast, shows decreased cephalic social capability and decreased response to eye contact and facial perception, perhaps due to a lower brain stem dysfunction.[27] There is considerable variation in musical ability among those with autism (table 6.2).[28]

When controlling for intelligence, autistic children often show demonstrable mechanical capabilities (although motor problems are co-morbid with some forms of autism),while at the same time a lack of social skills. They have decreased levels of oxytocin, and oxytocin infusion has some ameliorative therapeutic efficacy in autism.[29] It is thought, though untested, that Williams syndrome children may have higher levels of oxytocin. Autistic children often benefit from musical therapy, and perhaps this induces elevated levels of oxytocin.

Table 6.2

Various Features of Autistic and Williams Patients

	Autism	Williams
Sociability	*Low*	*High*
Musical engagement	Typically Low	High
Empathy	Low	High
Cerebral volume	Normal	Small
Paleocerebellar volume	Normal	Small
Neocerebellar volume	Small	Large

Source: Levitin 2005.

One treatment for autism is facilitating social engagement. One can set the conditions for such contact through music.[30] The issue is quite the opposite for individuals with Williams Syndrome, who are both musically and socially hyper-attentive. Children with Williams syndrome show a general decrease in brain volume.[31] But regions of the temporal lobe are actually greater in Williams syndrome than in controls, while the amygdala is decreased.[32] The amygdala of such children seems to be more reactive than controls to diverse social events.[33]

Preserved musical sensibility in Williams syndrome individuals is remarkable. Several studies have shown a greater liking of music in Williams syndrome individuals than age-matched controls.[34] Williams patients more readily engage in music than controls, while autistic patients show decreased perception of emotion in music.[35] The hypersocial feature overlaps with a tendency toward hyper musical engagement.[36] This engagement includes increased frequency in looking for music, playing music, and expressing emotional responses to music. A sensibility for and a sensitivity to sound seem to be features of these individuals.[37]

The temporal activation to music in controls versus Williams syndrome individuals demonstrates activation of the temporal gyrus and Heschl's gyrus, while also showing a more diverse and diffuse activation that includes the amygdala and cerebellum.[38]

Individuals with Williams syndrome have also been reported to have an expanded activation of the visual cortex. In a study using fMRI to measure brain activity, Williams syndrome individuals displayed greater visual cortex activation in response to music.[39] In addition, they showed diminished responses to anxiety associated with music.[40]

Music and Cognitive Adaptation

Neural disorders, such as Williams syndrome, tend to reinforce the assertion that music is an adaptation linked to diverse neural substrates, including motor regions tied to syntax and language. Music serves diverse communicative functions.

Humans come prepared with an arsenal of cognitive adaptations rooted in social discourse and commerce with one another and the construction of objects that we use: our tools. Our evolution is bound to social groups working in unison across diverse terrains. Key abilities also include recognizing the kinds of objects that are useful and avoidable, coupled with a wide array of inhibitory capacities that contribute to social cooperative behaviors.[41]

Neural expansion sets the stage for technological creations, expanding our capabilities. Tool use, for instance, and tool making were critical for this development by expanding our horizons of adaptation and by increasing comfort. The development of musical instruments demonstrates an advanced application of tool use.

Diverse cognitive adaptations, including our ability to predict the behaviors of others, are a function of the fact that we tag our fellow humans in terms of their beliefs and desires.[42] This, of course, is a higher order cognitive function. We use this adaptation to, in part, predict what other human beings will do in our social world based on our perception of their intentions.[43]

David Premack, one of my teachers in the 1970s, called this adaptation "theory of mind." The philosopher Daniel Dennett called something similar the "intentional stance." Both have to do with consideration of the beliefs and desires of others in predicting their behavior and coordinating one's own response. Many scientists in the early twentieth century were aware of the concept of theory of mind, but its placement in the cognitive language of prediction, with the consideration of what we are good at as biological creatures, reached a kind of formalization during the period of the cognitive revolution.

This ability to track others by what we think they desire and believe is an important cognitive resource. Of course, we track many behaviors that are simpler: for instance, joining eye contact on a common object in order to see what someone is looking at brings people together in a coordinated fashion. This type of behavior is at the heart of pedagogy. It is also essential for the foundation of music sensibility.

We learn from one another, manipulate one another, and predict behaviors by what the focus is on and what the eyes are rotating toward. This focus is externally and literally telling us something about the beliefs and desires of others.[44]

Music is a good vehicle to bond us to each other by producing social synchrony as well as musical harmony, in rhythm and in step with one another. Music allows people to play off one another and expand our horizons with others.

Children are oriented within the first few months of life to form social contact through the visual system and to track events in a manner of joint attention to objects.[45] These events act like social glue by facilitating future transactions with others and determining social adaptation. The social roots of our diverse cognitive capabilities are pervasive and, as I have indicated, are at the heart of music.[46]

Our evolutionary ascent is linked to our social ability, in addition to tool making and the onset of linguistic competences.[47] This is coupled with a long gestational period, a long lactation and dependency period, and with the massive amount of learning that takes place early in ontogeny.[48] In addition, there is a link between our longevity and the evolution of our problem solving capabilities. Our species had a greater opportunity to solve problems as humans lived longer. Our ability to transmit cultural knowledge and technological advances, of course, allows each generation to build on what has been discovered previously.[49]

"Motherese," the particular high-pitched nonstandard speech that adults, especially parents, use with small children, is a universal feature of the parent-child bond. Its sing-song pattern probably reaches deep into our human evolutionary past, bonding moments in which acoustical sing along is linked to diverse forms of gratification.[50] This leads to the production of endogenous opiates linked to social comfort in response to the mother-infant contact at the heart of social evolution in our species.[51] Song and motherese are both tied to olfactory sensibility, passed down in part through epigenetic mechanisms (that is, enhancing or silencing the genes for the production of hormones, for instance oxytocin or vasopressin).

Moreover, brain capability is linked to an expansion of knowledge about resources, where they are located, and other food-related behaviors such as tending to and cooking it.[52] Diverse foraging behavior and metabolism have been linked to an expansion of brain function and perhaps hominid evolutionary expansion.[53] Action and perception, including calls to others, are joined at the hip in these behaviors.

The brain is an expensive metabolic organ, and so a regulatory shift from the periphery to the central nervous system was an underlying adaptive factor for an expanding cortical mass.[54] An increased accessibility of brain stem function by cortical expansion is a basic blueprint of neural functioning in our species, cutting across a wide expansion of behavioral options.[55] Always, the degree of cognitive competence and social gesture, bipedal organization, communicative engagement, diverse tool use, and pedagogy are clearly linked to an expansion of the range of social contact.[56]

What have evolved in our species are long-term social bonds, plasticity of expression, an expanded cortex, and a wide range of behavioral functions. Social intelligence, particularly in primates, is importantly connected to reproductive success.[57] For instance, the alliances formed by female baboons and dolphins are vital to this.[58]

Primates are, for the most part, extensively social by necessity. Group formation is pivotal, and since so much learning is required for our species, longer periods of ontogeny are essential.[59] A brain that takes time to develop and a nervous system with neurons that form diverse networks across the cephalic systems in primates expands in scope; the development of the brain in ontogeny and prenatal development reflects the degree of cognitive and social competence that is necessary for humans.[60] Nature has thus selected physiological cognitive systems oriented to the social milieu. The evolution and expression of these systems underlies the diverse forms of complicated social assessments, making singing a form of social grooming.

Consider the complex social relationships of the orchestra leader: the diverse musicians, the strings, horns, flutes, drums, and so on, let alone opera, with its combination of drama and music, both instrumental and vocal. Even a moment of listening to a Verdi opera will provide a ready sense of the vastness of cortical expansion and social contact that inheres in this fundamental part of human expression: singing out to others and calling attention.[61]

Now imagine a musical improvisation: the complexity of the interactions, the scope of the music, and the underlying cognitive resources it must require. Those same resources are found, in part, in the regulation of social relationships, group size, hierarchy, distribution of food resources, shelter protection, dominance, and comfort through alliances. Such systems are quite varied and all involve cephalic innervations and expression. A premium is set on cognitive evolution; an expression of diverse cognitive/behavioral adaptations coupled with neural expansion.[62] Music is used, in part, as a sort of social adhesive.

Social and Musical Contact and Cortical Expansion

As I have indicated earlier, the greater the degree of social contact and social organization experienced by a human, the greater the trend toward cortical expansion.[63] In other words, group size and social contact are linked to neocortical expansion in hominids, as is longevity. The pressure of coming into contact with others, creating alliances, and tracking them no doubt required more cortical mass (see figure 6.3).[64]

Music is replete with social contact. In fact, its origins are in contact with others. Mothers making contact, calls to others, and

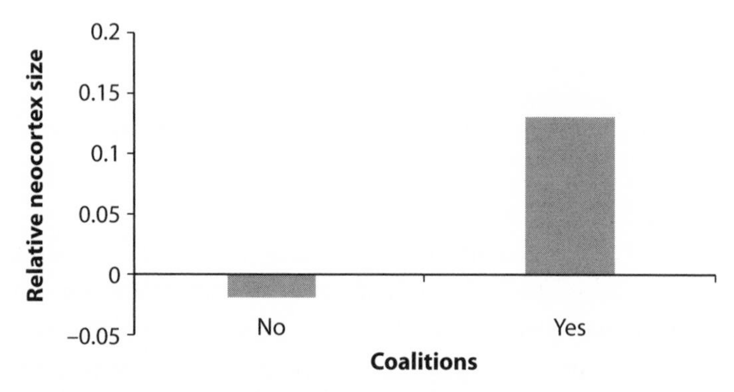

Figure 6.3 Neocortex size and social cooperation.
Source: Dunbar and Shultz 2007.

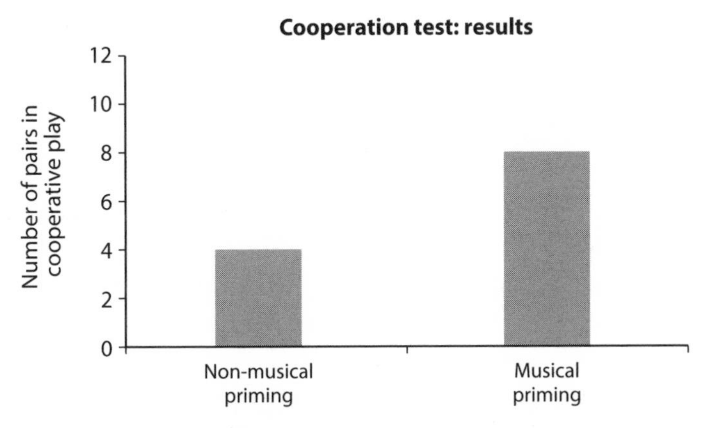

Figure 6.4 Joint social action and music making; musical priming leads to an increase in the number of cooperative players.

Adapted from: Kirschner and Tomasello 2010.

rhythmic patterns with others in the social group are all ways of keeping track of others, staying in touch with others, or playing with others. Indeed, exposure to music in young children is known to promote prosocial behavior in children. Studies in joint singing, or drumming, for instance, when controlling for diverse intellectual and personality factors, promotes prosocial behaviors (see figure 6.4).[65]

Interesting correlations have been suggested between neocortical size and social cognitive skills.[66] It is the expansion of cephalic functions that underlie the tool use that services physiological/behavioral regulation. An expanded cortical/motor system with diverse cognitive capacities is no doubt pivotal to our evolutionary ascent and to the musical instruments that we develop to facilitate social interaction. And thus, innovative tool use has been linked to an expanded cortical mass.[67]

What evolved in our species are long-term social bonds, plasticity of expression, and corticalization of function. As our cortical visual functions increased dramatically, standing up and forming eye contact began an evolutionary expansion in many primates, as well as the use of representations of objects and the use of tools as core cognitive capacities. Technology is an extension of ourselves that expands what we explore, including musical objects.[68] Innovative tool use, including the use of musical instruments, is not surprisingly linked in diverse primates to an expanded neocortex.[69]

A broad-based set of findings in non-primates has been the link between social complexity and larger brain size.[70] The metabolic investment of larger brained animals is expensive; neural tissue forms a high-energy organ. If brains expand, it probably means other tissue does not, or at least not to the same degree.

Two important pathways in the central nervous system underlie how we ascertain where an object may be located and what it may be.[71] This segmentation is tied to sounds and song.[72] Moreover, neurons in the premotor region, located within the frontal lobe, are contained to a large extent within Brodmann's area 6. This region is importantly involved in the direction of action and is linked to the larger organization of action, including that of musical expression, and the auditory influence in the organization of action.[73] Moreover, diverse regions of the temporal lobe have long been linked to social perception, eye gaze, and tracking the vector of visual systems of others, and would also underlie musical expression.[74]

Epigenetic Events and Social Contact

Music is tied to regions of the brain that mediate social sensibility.[75] Indeed, as Dewey long ago noted, lifelong learning and creative expression "of art as experience" provide a window into epigenetic expression, perhaps through regulation of the DNA genome.

Dewey, and later Meyer, emphasized the plasticity of our psychobiology; one feature is lifelong capability of learning. Music is one discipline; science is another. The difference between them is negligible in terms of creativity, experimental sensibility, and problem solving.[76] Music impacts the brain from neonatal exposure and throughout the life experience.

The regulation of oxytocin and social contact has been linked to an old controversy in biology. Epigenetics is still controversial, recalling debates between Lamark and Darwin about acquired traits and evolution,[77] but it has grown in acceptance with the integration of molecular biology into broader based and regulatory biology.[78] Epigenetics can be defined as changes in gene expression without altering the underlying DNA and is perhaps at the heart of phenotypic alterations in development contingent upon context and circumstance. Behavioral adaptations have long been noted to be fundamental to evolutionary change and indeed underlie any definition of evolutionary change.[79] One mechanism that may underlie epigenetic gene regulation is methylation and demethylation (e.g., silencing or enhancing the expression of genes, e.g., oyxtocin and vasopressin, on expression).[80] Demethylation prevents transcriptional expression, and one result is the silencing of gene expression. Another putative mechanism for silencing genes in development is via chromatin regulation.[81]

Variations in social behaviors, for instance, are linked to oxytocin (or prolactin, vasopressin) expression.[82] Higher levels of grooming in cross fostering experiments with primates are tied to estrogen and oxytocin expression, as well as glucocorticoid receptors and diverse neurotransmitter and neuropeptide expression such as corticotropin-releasing hormone (CRH), serotonin, dopamine, and DNA methylation.[83] In contemporary terms, research is showing both epigenetic and genetic contributions to group formation. For example, genes that produce oxytocin expression, an important peptide hormone that plays diverse regulatory roles from milk lactation to parturition, from social attachment to perhaps music expression, are particularly significant. The same hormone, thus depending upon where it is expressed in end organ systems, plays a wide variety of roles.[84]

Forms of musical therapy that act to ameliorate human discomfort and expose us to more pleasurable sensibilities are rife with form and function, perhaps via epigenetic regulation of neuropeptides such as oxytocin.[85] The embedded experiences speak to the diversity of expression across music and with common themes amidst a long, rich history. Neuropeptides such as oxytocin, tied to social attachment in both females and males, are vital for normal development and well-being. Oxytocin expression will be linked to this development, as will a number of other neuropeptides. The genes that underlie the expression of these information molecules will be regulated by epigenetic effects over our lifetimes.

Importantly, changes in demethylation are linked to the transmission of social behaviors to offspring during development in cross-fostering experiments. Indeed, demethylation impacts the transmission of social behavior and can be reversed later in life.[86] The imprinted genes are not permanent and the emphasis on silencing gene expression by imprinting or emphasizing is of course an active area of research. This is a general mechanism for the regulation of gene expression, impacting the function of oxytocin, vasopressin, or dopamine that underlies song. Vasopressin expression, for instance, is under the control of the gonadal steroid hormones (testosterone and estrogen). These events are mediated by epigenetic mechanisms via DNA promoter mechanisms, silencing or enhancing the expression of vasopressin.[87] Oxytocin might impact musical sensibilities over a lifetime by enhancing oxytocin gene expression in suitable musical contexts.

Conclusion

Musical expression begins early and plays a role throughout most of our lives. Cephalic capability is a pervasive feature, and the genes that set the conditions, along with the endless musical props that we hang onto, are embedded in, and are part of, our cultural evolution.

Some individuals are born with a natural talent to sing well, to play music well, and to display the excellence found in musical sensibilities. Music favors social and instrumental expression, problem solving, calling out to others, and marking places in space across extended time in cultural expressions of great diversity. A predilection to hear song, as hyper-manifested in Williams syndrome, is just that: a core predilection of the human condition.

Music is almost always tied to some social context: either the social self, or the individual on solitary walks imagining the many social parts of oneself, including the collection of expressions that constitutes one's history and experiences.[88] Cortical expansion is tied to social contact, and instrumental development underlies the expression of musical sensibilities.

Amidst a basic predilection to respond to music, to remember music, is the devolution of function that comes with age. Sacks, referring to a patient and former musician, says that, "At the piano she first blundered hitting the wrong notes and seemed anxious and confused."[89] But then, when she began to find her way with the Haydn piece, and began to play, "all is forgiven."[90]

CHAPTER 7

Music and Dance

The concept of dance might have been a new and startling innovation to the penguins in the movie *Happy Feet*, but song and movement go together in every human society: music to my feet, as it were. The human condition is linked to music and dance, and the range of human emotional expression is fundamental in this regard.

Susanne Langer was a professor of philosophy, during the middle of the twentieth century, at Connecticut College, a school with a significant dance program.[1] Many well-known modern dancers have contributed to and participated in this program (e.g., Martha Graham). Langer understood that movement and dance are at the heart of music, and music is at the heart of movement. She noted that "music must be swallowed by movement"; we are in movement, while dancing and singing.[2]

Indeed, movement and a sense of time are intimately connected, and the brain is prepared to detect movement, both familiar and unfamiliar. Our sense of self is tied to movement.

The concept of a river of movement and time also reflects our origins in water; after all, most of our body and the planet are comprised of water. Composer Claude Debussy's "Water Pieces," for instance, effectively captures the great flux of existence, of movement, of our feeling of being moved to sing, as well as our fantasies of singing and moving.

Langer has said that, "All the arts create forms to express the life of feeling;"[3] namely, cognitively rich feelings expanding in symbolic breath and reflective of the cultures in which we live.[4] Music is expressed across all cultures and has been with us, evolving for as long as we as a species have been here with instruments. Their complexity is a feature of the larger cultural milieu.[5]

Langer does not describe the expectation that permeates music, nor does she think of music as a sign or stimulus for something else. By contrast, Dewey and Meyer regard rhythmic generators as a feature of cognitive systems and cephalic function across music and virtually all human activities.[6] In other words, anticipatory expectations are endemic to cognitive function, musical or otherwise, and thus cognitive systems pervade all musical activities. Therefore, the distinction between the cognitive and the non-cognitive is undercut at all levels of the neural axis. Rhythmic expectations stand out when one considers the close link between music and dance. And indeed, music and dance mostly run together across cultures, pregnant with cognitive anticipatory expectations. Consider the George M. Cohan song "I'm a Yankee Doodle Dandy," now mentally inseparable from the dancing of James Cagney, a personal favorite of mine, capturing a very precise sense of place and history in dance and music.

Curt Sachs, a great historian of music and dance, noted three kinds of impulses tied to music:[7]

1. Motor, or what one might call a rhythmic motor generator
2. Ritual or structured behaviors
3. Melodic expression

In this chapter, I begin with the internal clock that regulates our sense of movement, and then continue the theme of art embedded in experience, music and dance being fundamental components of that experience. Dance and music are so linked that it is hard to listen to a rhythmic song without beating one's foot. Dancing is replete with musical sensibilities.

Movement, Clocks, and Cognitive Physiology

The dancing bees so elegantly described by Von Frisch, the twentieth-century German biologist, depicted the diverse uses of movement.[8]

The bee dance is highly functional, indicating the precise location of food sources. It is a semiotic act, rich in gesture for a bee brain; for some of us it is hard to observe and not feel its aesthetic component. Whether that is important to the bee or not is another question I cannot answer.

Cognitive systems bind us together and set the conditions for the organization of action amidst a world coherent but disruptive; regularities are pervasive. The twenty-four-hour rhythm is the essential rotation of our behavior and is tied to light/dark cycles. This clock is so essential that it is found in diverse regions of the body, such as the liver, as well as in regions of the brain.[9]

Clocks are linked to the prediction of periodicity as we cope with the world. The twenty-four-hour clock is just one of many: monthly, yearly, seasonal. There is also the time sense that can pervade our activities; whether we are aware of it or not, timing is endemic. Of course, timing mechanisms pervade music when we are keeping time, tempo, and the like. Time management and organization are also inherent in music, and entrained in aesthetics more generally. Indeed, entrainment to predictive core features is at the heart of nature and is tied to the anticipation of events. Endogenous clocks like the twenty-four-hour clock are used biologically to anticipate events, such as when food or other vital resources are available, and clock-like mechanisms underlie the tempo of music.[10]

Our many body clocks are pervasive in bodily rhythms, and clocks in every organ regulate diverse information molecules such as dopamine and vasopressin. But it is the cephalic clocks (e.g., twenty-four-hour clock) that are tied to music. These are the rhythms that we come prepared to perceive, entrain, and coordinate with, and with which we regulate ourselves in response to rhythmic patterns.[11]

For a colleague of mine who is a dancer, the rhythmic beat is at the heart of song; of course, he is not alone in this view. Naturally, the nervous system itself is rich in rhythmic function.[12] *Rhythms of the Brain*, neuroscientist Gyorgy Buzsaki's book, provides an excellent understanding of cognitive capability.[13] Inherent rhythmic properties ready the brain for action; there is relative rest, but not the elimination of the inherent patterns of the neural system that underlie all cellular and regulatory events. Indeed, evidence indicates that neural networks oscillate in functional networks independent of external stimuli.[14]

So, our bodies are full of all sorts of regulatory and innate clock-like rhythms, the most obvious one of which is the palpable beat of the heart, but most of which are unknown and undetected. In spite of this, they surely must figure into primitive rhythmic beating, sounds, and eventually songs. These clocks are tied to the rhythms and the core themes of nature as we forge coherence toward keeping time, tracking time, and organizing ourselves with others by "keeping to the beat" internally, socially, and ecologically.

The Greek concept of the "harmony of the spheres" is merely one expression of a sentiment that flows across diverse cultural experiences, and of the ecological rhythms of events, internalized in diverse end organ systems, which are not simply passive. When it comes to neurally generated rhythms, we become anticipatory and linked to others, something essential for the sense and evolution of both music and language.[15]

Human Activity, Art, Representations, and Human Inquiry

Dewey was fond of linking human activity with common threads: aesthetic experiences, problem solving, and forming a common currency. Art infuses life, and there is no higher or lower art. Folk music can be as elegant as Mozart, and as moving.

Indeed, the representation of objects is an ancient human activity, and we come prepared to understand objects in our environments. Cave art is but one example of this phenomenon that we find across cultures.[16] The desire to represent and understand the objects in our world is a fundamental human motivation. We also seek to orient ourselves to events and to objects, or predictions about events, as was discussed in chapters 4 and 5. This orientation is embedded in practice, in action,[17] in movement or dance, and in music.

The work of Doris Humphrey, a twentieth-century modern dancer, in *Rhyme*, for example, suggests that aesthetic sensibility evolved because elegant classificatory cognitive systems are adaptive.[18] We come prepared with a variety of cognitive systems, information processing, inferential mechanisms, and representational capacities for depicting and making sense of the objects that we are likely to encounter.[19] These systems pervade our experiences and help determine how we interpret and predict events. They also underlie our problem-solving proclivities that reach into aesthetic judgment.[20]

Such a predilection is wired in our minds/brains.[21] Aesthetic judgment is part of our appraisal of ourselves and of others, and is about both what is attractive and what we are trying to understand. It represents the objects we eat, copy, capture, and try to control.[22]

Kant tied aesthetic judgment to a "free play of the imagination," of cognitive capacities and faculties interacting to determine the perception of beauty or the sublime.[23] Kant and the free play of the imagination set the stage for the "mind to think [of] the unattainability of nature regarded as a presentation of ideas."[24] In Kant's lofty terms, the construction of aesthetics, musical or otherwise, pushes one beyond what can be represented and known toward a semblance of the sublime. This occurs through embodied rituals amidst an expanded sense of sensibility through play and practice, which pervades our sense of music and movement or dance.[25]

In Langer's view, information processing systems are continuous with bodily representation, and they reach their zenith in aesthetic sensibility or perception. As she stated, "All works of art create forms to express the life of feeling."[26] But *feelings*, I would add, are part of the perceptual apparatus that underlies information processing.

An aesthetic appreciation emerges from trying to capture nature amidst cognitive sensibility, tied to adaptation and aesthetic judgment, and intimately linked to bodily sensibility.[27] Perhaps music, often construed as one of the more abstract forms of representation, demonstrates this core cognitive capacity to reach beyond sheer beauty and the sensuous to the sublime and a free play of cognitive capabilities.

John Cage can play with silence, as Samuel Beckett writes in a play about endlessly waiting (*Waiting for Godot*), as a way perhaps of expanding our horizons in a form of "mindfulness," a term used to invoke thoughtfulness and expand cephalic capabilities. That is a reflection of our cultural evolution. Some sense of music is part of what is not present, and therefore has some presence.

The sense of aesthetics is a feature of the way we come prepared to interpret the world. Such aesthetics are historically variable and rich when the ecological conditions are suitable.[28] Aesthetic judgment reflects our cognitive flexibility, and our extension and use of specific cognitive mechanisms to widen domains of human expression.[29]

In experiments, Sarah, a chimpanzee with extensive communication training, was shown movement by humans that we would judge as graceful and aesthetically pleasing. She was then shown what we would consider a more awkward movement in comparison. The chimpanzee tended to look more at the more graceful performance.[30] Perhaps form matters to the visual systems of the chimpanzee. But these effects in Sarah are modest when compared to the aesthetic judgment of movement in a five-year-old human child. Children in the same experiment express a specific and forceful preference for the more graceful movement.

Bodily form, the reach for excellence and the appreciation of it, clearly serves biological functions (e.g., health, indication of good genes, of being able to master physical challenges). Aesthetic enjoyment of bodily shape is culturally variable but nonetheless reaches across cultures.[31]

Music and Dance

Music and dance are endlessly intertwined. Any history of dance is in part a history of music.[32] Diverse forms of life and ritual are expressed, and bodily rituals are a piece of nature, from the dancing of the bees to the synchronous bodily and howling rituals of wolves.[33]

Music and dance are linked not only in bodily expression but also in cephalic expansion. As Curt Sachs, the musical and dance historian, noted with regard to dance, "Only in the dance is the intrinsic union between man and animal consummated."[34] A noted music critic of the nineteenth century, Eduard Hanslick, stated, "the essence of music is sound in motion."[35]

Dance is a phylogenetically ancient expression in which no separation exists between our desire to move and our desire to reach into rich rituals of bodily expression. Ages in music, like ages in dance, are cognitive and historical constructs.[36] But we know how to link dance and ritual in ages past and present in the context of the larger cultural milieu.

For instance, within dance, a place in St. Petersburg was unique: an assembly called the Ballet Russes. There, Stravinsky, Fokine, and Ravel all produced music that broke boundaries, coupled with the elegance of movement of Nijinsky and Pavlova.[37] The combination of musical and dance talent produced an extraordinary explosion of creative energy. Many of the parties involved (e.g., Ravel) would later describe this extraordinary flowering of art.[38] While this assembly was unique in the breadth and quality of its output, these fusions are in fact commonplace and suggest the fundamental link between music and dance.

Martha Graham, for instance, changed the landscape of modern dance paradoxically with forms of ancient myths accompanying contemporary music (e.g., Aaron Copland). She produced a rich landscape of pieces, including *Lamentation*, *Primitive Mysteries*, *Clytemnestra*, *Appalachian Spring*, and *Dark Meadow*.[39]

Diverse composers, including one of my favorites, Aaron Copland, captured Americana music for Graham and for ballets. They were and are a wonder of breadth and depth. Copland was a senior colleague and early mentor of Leonard Meyer, who himself writes about rhythm, melody, and harmony, and whose work I have frequently discussed in this book. Copland himself produced a series of lectures at the New School for Research in New York City.[40] These lectures are remarkable in their prescience, and specifically note the link between music and dance.

A wonderful chain in dance runs from Isadora Duncan to Hanya Holm, and to Jose Limon and Alvin Ailey, capturing diverse spirits of movement, music, and meaning.[41] For Ailey, capturing the African American experiences of spirituality was central to his body of work, from pieces set to spirituals and church music, to the music of Ralph Vaughan Williams in "The Lark Ascending."

Music expands dance, and dance expands music: Both fuse as expanding landscapes of meaning. "Revelations," for instance, is a depiction of the dance and movements linked to black spiritual music and cries for freedom. Indeed, the choreography of Alvin Ailey called "Cry" conveys in movement a sense of story, toward some form of human liberation, a common theme.

At the same time, musical expression was also connected to themes of liberation; Marian Anderson's "Songs of Freedom" lionized that theme in song and expression in front of the Lincoln Memorial in 1939 after she was denied the right to perform in the Daughters of the Revolution Hall.[42]

Stories, Meaning, and Lived Experience

We are not Cartesian machines, thinking in an abstract divorced vacuum, nor are we random inductive machines. Musical sensibility makes this very transparent. We bring with us diverse forms of cognition that underlie what Dewey used to call "lived experiences," or what others have called "embodied cognition."[43]

Music is often full of purpose and bound to movement, whether literal or not.[44] This goes from the structured ballet of Chopin-type music, with its endless practice and the expansion of musical movements, through Charles Ives's "The Fourth of July." Music is expanded and tied to meaning, stories, and living experience running together in both long and short versions.

Our sense of self and our sense of music are often rooted in our life histories or our trajectory of movement through space. One such conception, adapted from Johnson, is depicted in figure 7.1.[45]

Thinking, movement, and music are understood in the context of action and transacting with others and are quite close to a pragmatist position where cognitive systems are embedded in the organization of action.[46] The emphasis is on embodied and expanded cognitive systems.[47] The sensorimotor systems are themselves knotted to cephalic machinations across all regions of the brain.[48]

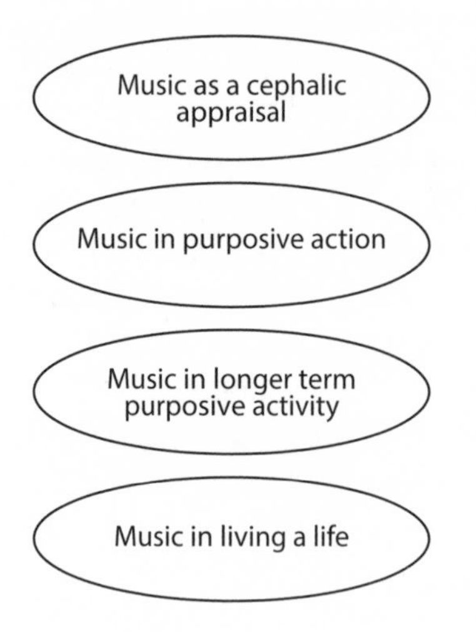

Figure 7.1 Phases of music.
Adapted from: Johnson 1993.

In other words, cognitive systems are, as I indicated earlier, not just a cortical affair, but endemic to cephalic function. Meyer, of course, understood this, with his kinesthetic/anticipatory description of musical sensibility. Movement and purpose, living a life with music rich in expectations and kinesthetic affordances, is replete with memory.

Indeed, Mark Johnson, a philosopher, and Steve Larson, a musicologist, have nicely depicted the common themes of music, movement, and musical trajectories and, consistent with the earlier works of Susanne Langer, the utter kinesthetic sensibility running through music.[49] Music is tied to movement and memory with vivid imagery, such as the one cited in "Over the Rainbow," and the expansion of our horizons by metaphors, as in George Harrison's "Something in the Way She Moves."[50] The metaphors of movement and destination traverse the "musical landscape" (see table 7.1).[51] Music is a force to be experienced.

The complicated and endlessly intricate movements of the dances of India are infused throughout with religious sensibility. These expanding categories of understanding, always toward the mystical and toward endless transcendental sensibility, are usually based on a narrative of travel, of being on a path.

An inherent tension in aesthetic sensibility, particularly in music, is the search for stability and stasis.[52] The construction of events

TABLE 7.1

Music and Metaphor

The "*MUSICAL LANDSCAPE*" Metaphor	
Physical Space (Source)	Musical Space (Target)
Traveler	Listener
Path traversed	Musical work
Traveler's present location	Present musical event
Path already traveled	Music already heard
Path in front of traveler	Music not yet heard
Segments of the path	Elements of musical form
Speed of traveler's motion	Tempo
The "*MOVING MUSIC*" Metaphor	
Source (Physical Motion)	Target (Music)
Physical object	Musical event
Physical motion	Musical motion
Speed of motion	Tempo
Location of observer	Present musical event
Objects in front of observer	Future musical events
Objects behind observer	Part musical events
Path of motion	Musical passage
Starting/ending point of motion	Beginning/end of passage
Temporary cessation of motion	Rest, caesura
Motion over same path again	Recapitulation, repeat
Physical forces (e.g., inertia, gravity, magnetism)	"Musical forces" (e.g., inertia, gravity, magnetism)

Source: Johnson and Larson 2003.

TABLE 7.2

Music and Others

Music and Others
Attribution of musical intention
Mirror others in musical worlds
Cooperative human interactions in music
Musical practices

shapes the way we orchestrate action in music and otherwise. The shared experience with one another is a fundamental adaptation,[53] and the cephalic innervations of bodily sensibility are a pervasive common element in our evolution.[54]

Our sense of moving through musical space or projecting ahead is influenced by experience and language. Of course, within this sense lies the effort of reading the minds of others, mirroring others, attributing intentional acts, and cooperating or not, all of which are features of music as a social medium of the human condition (see table 7.2).[55]

Perhaps not surprisingly, our sense of motion, movement, and physical sensibility impacts our sense of time and space. This means that there is a bodily component to our sense of time and memory of events, which is rich in sensorimotor experiences, as well as agency and action—and the experiences of music are boundless with their sense of movement.[56] As Eric Clarke notes, "Musical sound requires and inevitably involves movement."[57]

We come prepared with an evolved brain and a set of cognitive predilections that are situated toward context, flexibility, and perceptual embodiment about objects that are conceptually rich and vital to behavioral adaptation of action, perception, and the brain. The emphasis is on the adaptive nature of these systems that run through human cephalic capability, including that of two important expressions: music and dance.[58]

Dance and Music

Dance is infused with music, and music always conjures a sense of movement. They are joined at the hip, so to speak, and the diverse metaphors serve cognitive links as we extend the landscapes of music and dance.[59]

Labanotation (a standardized system of analyzing and recording human motion) is one representational system designed to capture and preserve movement and is analogous to notational systems in music, preserving previously ephemeral pieces so that others can recapture them.[60] Of course, in our era of multiple media outlets, a dance can always be captured in visual media. Interpretation is important to our understanding and appreciation of music and movement.

Gesture fills the space of Fred Astaire and Ginger Rogers, leaping with ease so gracefully, as is immediately called to mind by those who know their work. Fred and Ginger exemplify how we are all physical in our singing, how our bodies swing to and fro to music.[61]

As Zbikowski notes, "The connection between dance and music seems completely undeniable—Astaire and Roger's movements are simply the fanfare from Kern's tune."[62] The gestures are brimming, bristling, and breezing, effortless and beautiful (see table 7.3). The music is highlighted by the dance and the dance is highlighted by the music. The dance completes the music, and the music is embedded in the dance. The music gains something from the dance, of course, as well, but it seems like dance may not really exist without music, in the sense that we are talking of it now.

Cognitive/motor systems of the sort alluded to throughout this book pervade the posture of Astaire at the piano playing a song like "Someday." Music and movement run together, just as cognitive systems are grounded in human action, understanding, and what Meyer, following Dewey, understood as "embodied," a term now very much in vogue, and rightfully so. Again, there is no Cartesian separation of a mind in a body in our actual experience.

Fred and Ginger exhibit sheer elegance and beauty in telling a story, an old story in the heart of the Depression, brimming with

TABLE 7.3

Musicial Gesture of Fred Astaire in *Swing Time*

Line	Lyrics	Gesture Feature
1	Someday	dip of head
2	When I'm awfully low	circular movement
3	And the world is cold,	circular movement
4	I will feel a glow just thinking of	circular movement
5	You	dip of head
6	And the way you look tonight	movement away from and back toward keyboard, dip of head

Source: Zbikowski 2011a.

possibilities for an American culture ripe for these tales. The music and motion filled the hunger in the Great Depression for song, dance, and something filled with possibilities.

The cognitive architecture is in part revealed in action; the dance and the song revealing gestures of cognitive structure.[63] Astaire, in one biography, was quoted as saying that, "Gershwin wrote for the feet"—though Astaire could carry a tune, as well.[64]

The stately Bill Robinson, an elegant man with endless dignity and grace, danced with the little sweetheart of America, Shirley Temple, again in the grips of the Great Depression.[65] African Americans in many parts of the country were not supposed to touch or mingle with whites in this manner, but there they are, moving through space, up and down a stairwell with elegant grace and genius, on cue together.

The improvisational genius of a Louis Armstrong, and those lucky enough to have been around to learn from him (e.g., Benny Goodman), required expanded cephalic capabilities. This meant a blending of the social and the motor, inherent in the communicative genius along with the musical genius of near perfection.

"Musing with the Gods"

Musement is at the heart of hypothesis formation. As C. S. Peirce, the philosopher/scientist noted, it has something to do with capturing the rhythm, being part of it—music and movement. Music is tied to the organization of movement in cognitive systems within the motor system, linked to different kinds of information processing within the brain. Premotor planning and motor control are inherent in this cephalic process between the order of behavior (or what Peirce called codified habits or frozen mind) and expressed syntactical patterns of movements that are the products of generative processes.[66] We distinguish between animate and less animate movement. Animacy is a core category in our cognitive arsenal for understanding the world.

These categories set the conditions for coherent action in adapting to terrains both social and ecological. These categories are not abstract, and in fact, they are set in everyday action. A neuroscientific perspective on dance ties diverse regions of the basal ganglia (e.g., putamen) and diverse regions of the cortex, including the premotor and motor cortex, and the cerebellum in the organization of dance.[67]

Part of our evolutionary lure is to expand the use of categories, such as agency or animacy, beyond perhaps where they are legitimate.[68] They are then instantiated in diverse forms of human activity, including musical sensibility. In addition, the mechanisms that foster social cooperative behaviors are tied to musical sensibility. Our respect for individual expression and initiative, and for a narrative and metaphorical expression of categories endlessly embedded in our stories, reflects an evolutionary predilection and adaptation.

A predilection to expand and attribute concepts such as agency has been extended to the gods.[69] The link between finite agents and a conception of infinite agency is an easy step from an adaptation to understand others in terms of movement and animate agency.[70] These spiritual sensibilities are often embedded in music and dance as in, for instance, India. Or consider, for instance, the cognitive categories involved in the movement of the Whirling Dervishes.

Dopamine would underlie every part of that ceremonial expression, from the sense of self-transcendence to the social contact knotted to brain regions involved in interceptive processing.

Expanding predilection in activity is a core feature in our creative sensibility in reaching to aesthetic sensibility and perhaps the "neural sublime":[71] a capacity to search for the unbounded amidst the finite and bounded.[72] The poetry or the odes and rhythmic sensibility inherent musically in Greek or Indian tales, to name just two, reflect the extension of cephalic categories of understanding and representing.[73] The animate categories pervade the poetic seduction.

Music is infused with the extension of the categories of animacy and agency, two core categories that underlie our sense of purpose, action, direction, and meaning. Embedded in the narratives, of which poetry is a rhythmic example, is the rumination of time as in "Burnt Norton" by T. S. Eliot, in which the directions of time and the "words move, music moves" in time.[74]

Poetry, like music, is less literal in general than language; of course, language nevertheless is overflowing with metaphors that expand our conceptual horizons.[75] Poetic metaphor is infused in the dances and movements and the stories. As Doris Humphrey noted, "Rhythm so permeates every aspect of a human being and indeed of the known world that it might be compared to the ambience of existence, like the water in which the fish moves."[76]

Religious dance, as diverse as the Shakers and Whirling Dervishes, expresses the ecstasy of movement and the simplicity of aesthetic ascendance through movement and dance. The song matches the movement.

The diversity of such expressions is as rich as languages are, and indeed, that is one of the cardinal features of the universal expression of song and dance. The religious sense is one dominant feature. Animacy and agency are two cognitive categories that figure importantly into the organization of movement, music, and in the organization of action—in this case, an action linked to religious expression in song and dance.[77]

Conclusion

Dance is tied to our sociability. Without the ability to move together, music becomes far less social. It is the sensation of "feeling the music" that really brings people together around music.

Music and dance co-evolved in contexts of adaptation, human meaning, and social contact. Within this is a mixture of what Meyer called "an aesthetics of stability," a coherence of expansion into the unfamiliar and the expected.[78] Metaphors within music expand our sensibilities. Metaphor and myth run through musical thought, as they do in all of human thought, through expression of mythic belief codified in movement and music.[79] Much of the infusion of creativity in music and dance, and everything in the Ballet Russes, for instance, took place amidst a moment of broader cultural change and uncertainty. The aesthetic ascended to the sublime.

But, even the less sublime, everyday manifestations of "singing in the rain, what a glorious day," show us that music and dance are a piece of our biology and an evolving part of our culture.[80] What pervades is the recurrent familiar amidst pockets of change and variation.

From Gregorian chant to Gospel music, to the organ in a church, to Davening, music and the spiritual are endlessly bound together, and indeed co-evolved together by weaving tapestries of memory and performance. Awe and the spiritual permeate musical expression.

Conclusion

Music and Well-Being

We are a social species. Vocal capability expanded that fact in the form of social competence. Music reinforced our social instincts, and then, coupled with the development of instruments for touching each other through sound, our social expression continued to swell. With the evolution of culture and memory coded into external artifacts, music became a core resource.

Music evolved in the context of social contact and meaning. Music allows us to reach out to others and expand our human experience toward and with others. This process began with song and was expanded through instruments and dance.

A series of steps set the conditions for this core capability in our species. A change in the vocal apparatus, leading to a larynx of a certain size, shape, and flexibility, is but one example. A vocal capability tied to social awareness along with other cephalic capabilities, converged together in behavioral coherence. From parental contact to the acquisition of food resources, music inheres across the cephalic and extends bodily expression toward others.

The evolutionary record suggests that musical instruments were perhaps well expressed over fifty thousand years ago in simple flutes and pipes and were depicted in our art (e.g., on bison horns).[1] What began as an extension of communication in a social context became something greater, something that was enjoyable in itself. Our evolution is tightly bound to music and to the body as an instrument (e.g., clapping). Music serves, among other things; to facilitate social cooperative and coordinated behaviors—the induction of "social harmonies."[2]

As Dewey noted, art is part of the ordinary human experience and is part of the framework in which our worlds are understood, adapted, and invented. Music is just one human activity among others in which expectations are cardinal features of the organization of human action and capability.[3] Musical expertise forms patterns of expectations, categorical markers, and sentiments amplified by depth and beauty.[4]

Folk songs, present across the planet, are captured in text and in music by Béla Bartók (see his Hungarian explosion in prose and music) and many others. Folk anthropology is a celebration of human expression, something that came to be in one of the humanistic impulses that surfaces periodically in our history. The human quest for understanding and self-expression is highlighted in music, and within this tradition is part of the breaking of barriers that divide us. Of course, music is not always blind; the nationalism and myopia that can predominate have always had musical expression. The horrors that motivate us to kill each other are emboldened by diverse folk cultural expressions (the song celebrating Hitler Youth from the musical *Cabaret*, "Tomorrow Belongs to Me," is all the more chilling because of its inherent, simple beauty).

An experimental sensibility broadens the horizon as we go from the familiar to the novel. For instance, the Fourth of July piece by Charles Ives stretches from one extreme to the other, both breaking and maintaining musical continuity.[5] Our sense of continuity has its roots in problem solving and in making sense of discrepancies. Memory and the diverse forms of cultural expressions provide a rich array to draw upon in the expansion of musical expression as we explore the unfamiliar amid the familiar.[6]

Our ability to link imagination to tools and instruments is a feature of our evolution. The organization of syntax in Broca's area and other cognitive/motor regions (e.g., basal ganglia) in our brain is accessed by the auditory systems for music and language, but also as part of the "serial order" of more general behaviors (see figure C.1). Moreover, diverse neural systems are recruited in "tracking" the musical form of action.[7] That is, rhythmic patterns inherent in tissue connect cohesively, both internally and externally with others.

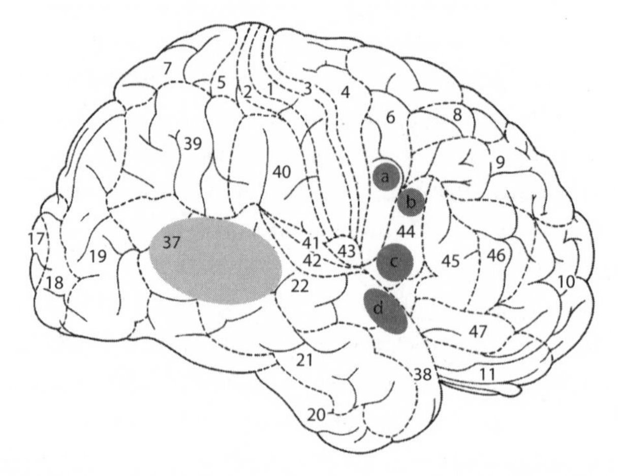

Figure C.1 Spatial aspects of processing syntax and semantics in music. The four smaller shaded areas represent activation foci described in previous imaging studies on music-syntatic processing: (a) vIPMC, IFLC ([b] superior and [c] inferior pars opercularis) and (d) anterior STG. The large shaded area (37) represents an area that is assumed to be involved in the processing of musical semantics, and in the integration of semantic and syntactic information, which includes parts of Broca's Area.

Source: Reprinted from *Current Opinion in Neurobiology*, 15:2, Koelsch, Stefan, "Neural substrates of processing syntax and semantics in music," pp. 207–12, copyright 2005, with permission from Elsevier.

Musical sensibility is a panoply of emotions. These emotions are inextricably linked to our cognitive, motor, and premotor resources and are expressed in everything we do, most especially in music. We easily scaffold to the social milieu in which we enact our musical historical circumstances with minimal effort.[8] Music is well ingrained and easily expressed in cultures, and is rich in style and function.

Stefan Koelsch proposes a recurring but rather bold hypothesis "that the human brain, particularly at an early age, does not treat language and music as strictly separate domains, but rather treats language as a special case of music."[9] This is a tradition that can be traced to the thought of Rousseau, and placed in a context of adaptation in which song and speech, or language use, are coextensive in the brain. Indeed, song and language syntax overlap to a great degree, including semantically; impairments in one are often linked to impairments in the other.[10] Indeed, the separation between the two is quite permeable. Both provide a vehicle for the child to gain a foothold in the world around her and to gain familiarity with objects that matter or those that are dangerous.

My own view is close to what Ian Cross suggests, namely that music and language enhance each other with regard to cephalic function and behavioral adaptation, jointly expanding our capabilities.[11] But then, language digs into everything eventually, and is quite pervasive. Like language, music is essentially rooted in social contact. But in spite of the social nature of music, its application in solitude also plays an important role in human existence and human expansion. These solitary moments are often the beginnings of great creations.[12]

Singing in the shower, singing in the rain, dancing in the rain: we have a lot of alone time for which music is a companion. A wide array of human pleasures underlies musical experience.[13] There are diverse regions of the brain that underlie syntactical and semantic structure. Music and speech vary with culture and are reflective of tonality within both auditory and non-auditory regions of the brain.[14]

The experience of music itself reflects the interaction between the performer and the person listening.[15] We come prepared to recognize the probabilities of an event, or what Huron, echoing Meyer, has called "sweet anticipation."[16] Cognitive expectations about music are fundamental to our understanding of it.

Dewey emphasized human growth within a behavioral context of diverse forms of appetitive and consummatory experiences that underlie all human activity, for which, as Meyer stated, "Both music and life are experienced as dynamic processes of growth and decay, activity and rest and tension and release."[17] This view is at the heart of the pragmatist notion of human development and inquiry, musical and otherwise.[18] In music, "The tension and struggle has its gatherings of energy and discharges, its attacks and defenses, its mighty warrings and its peaceful meetings, its resistance and resolutions, and out of these music weaves its web."[19] The rich semiotics that permeate music, despite the lack of details of reference associated with language and other conceptual abilities, is a key factor. Musical aesthetic imagination is a wonder, an awesome sensibility, which intensifies the human experience of the world around us.

The liking of music is one the great gifts of cephalic predilection in our species, and probably in other animals, both extant and extinct.[20] Music in its origins is about social communicative competence, extension toward and contact with others. Song and the building of instruments were used to mobilize life-building actions.

Biophilia is the love of life, necrophilia that of death. Both find emotional expression in music. But the basic predilection for music is found in a love of life and adaptation, reaching out beyond oneself. A core instinctive response to this is found in a sense of our well-being.

Erich Fromm notes, "Biophilia is the passionate love of life and of all that is alive; it is the wish for further growth."[21] Music is an integral part of life, and without it there is a loss of meaning, pleasure, and core human expression. Music and our sense of well being starts early in ontogeny in the formation of social contact and is an endless source of gratification throughout life. The connections between music and our way of living and love of life illustrate the fact that the devolution of musical function is a decrement in human well-being.

Oliver Sacks, a neurologist with a real feel for the patient and also a flair with the linguistic keyboard, writes about musicophilia. He suggests that, "Perhaps musicophilia is a form of biophilia, since music itself almost feels like a living thing."[22] One view of musical expression and artistic expression in general is that they are love affairs of the self toward the creation of objects, or musical pieces prior to musical instruments.

Nietzsche and Freud both notably captured the destructive dark side of human aggression. The expression of natural aggression can easily descend into wanton destruction toward others. As the noted etholologist Konrad Lorenz pointed out, only our species is so successful at excessively destroying life. Other species can and do display diverse forms of aggression, but it is nothing like our species' ability to, for instance, torture or wage war.

The machinery for aggression and destructive behaviors is omnipresent, and music emerges throughout. Music is an important form of human expression of all aspects of human feeling. For instance, abject endless wandering (for Mahler), or Wagner's reach and height for the eternal destiny, stand as examples. Even war calls may be considered a form of music, and the beat of a drum driving one to battle is an omnipresent theme in war imagery. The interaction between life affirmation and destructiveness is in constant tension. The experience itself is primordial. Like all things human, music is also easily, all too easily, abused; the harmonies of group think can also be devastating.

A phrase of Nietzsche's, "human all too human," expresses with force this dualistic feature of music. Music is related to the Dionysian, and the freedom and power associated with this Greek God: primordial, generative, and destructive.[23]

In *The Birth of Tragedy,* Nietzsche speaks about how humans experience music in "their mother's womb," by which the world is understood "through unconscious music relations."[24] Indeed, one could consider that deprivation of music could be a kind of life without meaning. For Nietzsche, like his intellectual mentor Arthur Schopenhauer, music is an ultimate expression of cephalic capabilities and an ultimate human wellspring, because "music is the will itself."[25]

Germany has harbored a vast array of great musical expression, mind boggling in its cultural expression. These expressions include the fugues of Bach, the mathematical formal beauty and endless complexity and depth of Beethoven, and the sheer brilliance of Mozart (along with his sense of fairness in the drama and his pieces that are romantic and innovative in expression of the expected and unexpected).[26] The cultural milieu was captured in music, song, and drama during a very small period of time.

Wagner, an unwise man, nonetheless wrote music that is incredible, clear, and represents the culmination of the romantic form of music in Western Europe: its heroic conception of love.[27] Loss and fate are elevated to their highest pitch in Wagnerian music.[28]

Wagner's romanticism expands the heights through clear expression in the music, with a striking sense of power and passion. Nietzsche warns us about Wagner, however, and in his *Autobiographical Fragments* reminds us that, "God has given us music so that it might lead us upwards."[29] Wagner's music, of course, was used as a tool for those leading his nation into the destructive culture of death that was Nazism.

Ultimately music is like dopamine: neutral, ultimately, in how it mobilizes behavior and human experience. It can be used for good or for destroying others. We march to war under many stripes and codes of musical expression. When we experience the heights of love, we express them most vividly in music. Music is part of the motivation to persevere, to mobilize, but also to attack. Just as language is neutral, but can be ill advised and render bestial expression more likely, so too with music.

What we know, though, is that music often mobilizes us toward action. I once listened to a Holocaust survivor who was over one hundred years old describe playing music in the concentration camp, surviving the experience, and playing music now. Clearly a musicophiliast in the extreme, she said, "Music is God," a phrase reminiscent of Nietzsche's, and a view shared by many people. This shared sentiment brings many of us together across boundaries.

This is mere speculation, but perhaps, as discussed in chapter 2, the neuropeptides, oxytocin and vasopressin, are activated to sustain broad social contact and a sense of well-being amidst the trauma of adversity, surviving, and continuing over the course of one's life.

The broad array of emotions tied to social contact and well-being are also tied to cephalic development and maintenance over our lives.[30] Their origins are tied to our evolutionary ascent as social animals and our instrumental development along with our musical sensibilities.

Music permeates the brain as a core feature, from pitch and rhythm to tempo and affect. The melodies dance across our brain, memory guides them through our lives, and the tension and release, or resolution, form an outstanding aspect of the experience of many forms of music and neural processing of events.[31] Perhaps not surprisingly, the amygdala, a region intimately linked to expectation and social contact, is also linked to improvisation of musical expression.[32] A small amount of uncertainty is a titillation of desire, something that initiates the exploration of the known and the unknown, something that infuses musical experiences with meaning.

Notes

Introduction

1. Gibson 1966, 1979; Clark, A. 1998; Clarke, E. 2008.
2. Dewey 1925/1989.
3. Berridge 2004: 179–209.
4. Dewey 1896; Meyer 1967; Galison 1988; Shweder 1991; Schulkin; Premack 1990.
5. Jaspers 1913/1997.
6. Chomsky 1965; Miller et al. 1960.
7. Martin 2007; Barsalou 2003.
8. Gibson 1979.
9. Clarke 2008; Donald 2001, 2004.
10. Clarke, E. 2008; Davies 1994.
11. Hatten, 2004; Clarke 2008; Clarke and Cook 2004.
12. Langer 1937; Clarke, E. 2008.
13. Donald 1991; Noe 2004; Clarke, E. 2008.
14. Clark, A. 1998.
15. Donald 1991. See also Lashley 1951.
16. Peretz and Zattore 2003; Pfordresher and Brown 2009; Stewart 2005.
17. Pfordresher 2006; Deutsch 1999.
18. Rozin 1976, 1998.
19. Meyer 1967, p. 11.
20. Meyer 1956.
21. Peirce 1880.
22. Rosner and Meyer 1986.
23. Meyer 1956.
24. Meyer 1956, 1967.
25. Ibid.
26. Jackendoff and Lerdahl 2006.
27. Dewey 1896; Berylne 1954, 1966; Loewenstein 1994.
28. Quote from Sacks 2008, p. 72.
29. James Fennimore Cooper (1838, 1852), *Home as Found.* New York: Stringer and Townsend, pp. 139, 214.
30. George and Ira Gershwin, *Girl Crazy.*
31. Zbikowski 2005.
32. Rozin 1976, 1998.
33. Gallistel 1980; Schultz 1932/1967).
34. Dewey 1894.
35. Peirce 1878; Dewey 1896; Berthoz 2002; Schulkin 2000, 2007; Barton 2004.

36. Lakoff and Johnson 1999.
37. Jackson 1884/1958; Decety et al. 1994; Jeannerod 1997.
38. Rozin 1976; Sloboda 2000, 2005; Chomsky 1965; Pinker 1994.
39. Helmholtz 1873; Temperley 2001. See also Rameau (1722) for an early exposition on harmony and musical expression.
40. Helmholtz 1873; Meulder 2010.
41. Meulder 2010.
42. Decety et al. 1994; Passingham 2008.
43. Hari et al. 1998.
44. Rizzolati and Arbib 1998.
45. Rauscheker and Scott 2009.
46. Calvert et al. 1997.
47. Zatorre 2001.
48. Shepard and Cooper 1982.
49. Farah 1984; Kosslyn 1980.
50. Kosslyn 1986.
51. Zattore et al. 2002; Zatorre and Halpern 2005.
52. Zatorre 2001; Zattore et al. 2002.
53. Sullivan, J.W.N. 1927/1955; Thayer 1871/1923/1973; Robert Neville (1981) personal communication.
54. Sacks 2008.
55. Swanson 2003.
56. Schultz 2007.
57. Berridge 2006; Schulkin 2007.
58. McGaugh 2003.
59. Herbert and Schulkin 2002.
60. Mithen 2006.
61. Rousseau 1966; Whitehead 1938/1967.
62. Cross 2010; Cross and Morley 2008.
63. Meyer 1956; Sloboda 1985/2000, 1991; Huron 2008.
64. Cross 2010.
65. Brown 2003.
66. Diderot 1755/1964; Rousseau 1966.
67. Cooke 1959/1964.
68. Rousseau 1966.
69. Nietzche 1871/1927.
70. Ibid.; Schopenhauer 1844/1966.
71. Adorno 1976.
72. Barzun 1949.
73. Darwin 1872/1998.
74. Temperley 2001; Jackendoff and Lehrdahl 2006.
75. Peirce 1903–1912/1977; Meyer 1956; Langer 1972; Myers 1905.
76. Sachs 2008; Myers 1905; Cross 2010.
77. Brown et al. 2006.
78. Cross 2010; Temperley 2001; Sloboda 1985/2000; Huron 2008.
79. See Brown 2009.

80. Huron 2001; Cross 1999.
81. Kant 1792/1951.
82. Dewey 1934/1958.

Chapter 1: Music and the Brain: An Evolutionary Context

1. Humphrey 1976.
2. Marler 1961, 2000; Smith 1978; Cheney and Sefarth 1990, 2007; Corballis and Lea 1999.
3. Juslin and Vastfall 2008; Juslin and Sloboda 2001.
4. Herman et al. 2007.
5. Tomasello and Carpenter 2007; Kagan 1984.
6. Dobzhansky 1962.
7. Barton 2006; Schulkin 2009.
8. Zatorre, et al. 2007; Peretz and Zatorre 2003.
9. Mithen 1996.
10. Dewey, 1934; Gibsonand Ingold 1993.
11. Mellars 1996, 2004.
12. Blake and Cross. 2008.
13. Cross 2009a; Cross and Morley.
14. Donald 1998.
15. Kripke 1980.
16. Gallistel 1990.
17. Leslie, et al. 2008.
18. Lakoff and Johnson 1999.
19. Rozin 1976.
20. Bordman 1992; Suskin 2010.
21. Gallistel et al. 2005; Spelke 1990; Dehaene 1997.
22. Darwin 1871/1874.
23. Gould and Eldridge 1977; Wood 2000; Foley 1996.
24. Mellars 2006a, b.
25. Lieberman and McCarthy 2007; Mellars 2004.
26. Boyd and Richerson 2005; Mellars 2006a.
27. Asfar et al. 1999.
28. Mithen 2006.
29. Roth and Dicke 2005.
30. Falk 1983.
31. Barton 2004.
32. Helmholtz 1873.
33. Rauschecker 2006.
34. Zatorre and Halpern 2005.
35. Kraus and Chandrasekaran 2010; Wong et al. 2007.
36. Rozin 1998.
37. Dowling and Harwood 1986; Dowling 2002.
38. Hebb 1949; Rauschecker and Korte 1993.
39. Hamilton et al. 2004.
40. Peretz and Zatorre 2003.

41. Perrett and Emery 1994; Rolls and Treves 1998; Adolphs 1999; Schulkin 2009.
42. Landau and Hoffman 2005.
43. Rauschecker and Korte 1993.
44. Weeks et al. 2000.
45. Lieberman 1984.
46. Fitch 2005; Rozin 1998; Lieberman 1984.
47. Lieberman 2011, p. 317.
48. Fitch 2005.
49. Lieberman 1984, 2002.
50. Lahr and Foley 1998; Mellars 2006a, b; Gamble et al. 2004; Mayhew et al. 2008; Finlayson 2004; Conard 2009.
51. Tattersall 1993.
52. Conrad 2009; Mellars 2006a, b.
53. Noonan et al. 2006; Green et al. 2006.
54. Mithen 2006.
55. Krause et al. 2007.
56. Lai et al. 2001.
57. Enard et al. 2009.
58. Krause et al. 2007.
59. Mithen 2006.
60. Ibid.; Mellars 2004.
61. Mithen 1996, 2006.
62. Mithen 2006.
63. See Morley 2003, 2002; Cross and Morley 2008.
64. Conrad et al. 2009.
65. Gunz, et al. 2010.
66. Rozin 1998.
67. Humboldt 1836/1971; Chomsky 1965; Pinker 1994; Lieberman 1984.
68. Rozin 1998.
69. Lieberman 1984.
70. Rozin 1998.
71. Wynn and Coolidge 2008.
72. Juslin and Sloboda 2001; Cross 2009B; Morley 2002.
73. Rousseau 1966, p. 136.
74. Mithen 1996, 2009.
75. Smith 1977.
76. Cross 2009b.
77. Grauer 2006; Nettl 2006.
78. Brown et al. 2004; Merker 2005.

Chapter 2: Bird Brains, Social Contact, and Song

1. Marler and Hamilton 1966; Smith 1977; Houser and Kloesel 1992/1998; Houser et al. 1998.
2. Marler and Hamilton 1966.
3. Mithen 2006.

4. Dunbar 2003.
5. Wingfield et al. 1999.
6. Strand 1999.
7. Wehr et al. 1993; Strand 1999.
8. Nelson et al. 2002.
9. Oftedal 2002.
10. Fitzsimons 1999; Denton 1982; Manzon 2002.
11. Fitzsimons 1979, 1999.
12. Strand 1999; Gimpl and Fahrenholz 2001.
13. Strand 1999.
14. Kanwal and Ehret 2006.
15. Ibid.; Insel 2010; Veenema and Neumann 2008.
16. Schulkin 1999; Smith, G. P. 1997.
17. Herbert and Schulkin 2002; Moore 1992; Insel 2010.
18. Phoenix et al. 1959; Goy and McEwen 1980.
19. Power and Schulkin 2005.
20. Keverne and Curley 2004; Carter 2007.
21. Choleris et al. 2007.
22. Kelley 2004.
23. Ibid.
24. Ibid.
25. Ibid.
26. Marler et al. 1995.
27. Marler et al. 1999.
28. Goodson 2008; Goodson and Bass 2000; Young 1998.
29. Goodson and Bass 2000.
30. Kelley 2004.
31. Goodson and Bass 2000.
32. Marler 2000.
33. Peirce, 1903–1912/1977; Marler 1961.
34. Nice 1943.
35. Thorpe 1974; Marler and Hamilton 1966.
36. Thorpe 1974; Armstrong 1965.
37. Cross 2009; Morley 2002.
38. Hartshorne 1973.
39. Dewey 1925/1989.
40. Hartshorne 1973.
41. Craig 1918; Tinbergen 1951.
42. See chapters 3, 4, and 5.
43. Aristotle 1968; Atran 1990/1996.
44. Atran 1990/1996.
45. Gurney and Konishi 1980; Gurney 1982.
46. Marler 1988; Liu et al. 2004.
47. Marler 2000.
48. Nottebohm 2005.
49. Ibid.

50. Marler and Pickert 1984; Garamszegi et al. 2007; Chomsky 1972.
51. Brenowitz 1991.
52. Marler 1981.
53. Nottebohm 2005.
54. Ball and Balthazart 2002.
55. Gurney 1982.
56. Arnold 2002; Marler 1961, 2000.
57. Gurney and Konishi 1980; Brenowitz 1991; Gurney 1982.
58. Marler and Doupe 2000; Gurney and Konishi 1980; Brenowitz 1991.
59. Ming and Song 2005.
60. Rosenzweig 1984.
61. McGaugh 2003.
62. Shors et al. 2001; Gould et al. 1999.
63. Gould et al. 1999; Shors et al. 2001.
64. Kempermann 2006.
65. McEwen 1995; McCarthy 2008.
66. Marler et al. 1988; Hauser and Konishi 1999; Marler 2000.
67. Arnold 2002; Marler 1961, 2000.
68. Nottebohm 1994; Thorpe 1974; Marler 1961, 2000.
69. Marler 2000; Fortune et al. 2011.
70. Marler et al. 1988; Nottebohm 1994; Schwabl 1993.
71. Voorhuis et al. 1991.
72. Bentley 1982.
73. Strand 1999.
74. De Vries et al. 1986; De Vries and Simerly 2002.
75. Fitzsimons 1979.
76. De Vries et al. 1995.
77. Goodson 2008.
78. Kelley 2002s.
79. Crews 2005, 2008.
80. Ball and Balthazart 2002.
81. De Vries 2008; Wang et al. 1997; Wang and De Vries 1993; Wang 1995.
82. De Vries et al. 1986; De Vries and Simerly 2002.
83. Carter et al. 1997/1999; Carter and Getz 1993; Donaldson et al. 2008.
84. Hammock and Young 2004; Donaldson and Young 2008.
85. Nephew and Bridges 2008.
86. Thompson et al. 2006.
87. Voorhuis et al. 1991.
88. Ibid.
89. Maney et al. 1997.
90. Leung et al. 2009.
91. Goodson and Bass 2000.
92. Keverne and Curley 2004.
93. Newman 2002; Schulkin 1999.
94. Insel and Young 2001; Goodson and Bass 2000.
95. Moore and Rose 2002.

96. Veenema and Neumann 2008.
97. Donaldson and Young 2008.
98. Gimpl and Fahrenholz 2001; Unväs-Moberg, K. 1998; Carter et al. 1997/1999.
99. Bowlby 1988; Hofer 1973; Grossman and Grossman 2003.
100. Herbert and Schulkin 2002.
101. Insel 2010.
102. Kanwal and Ehret 2006.
103. Adolphs et al. 2001; Zatorre 2001.
104. Gigerenzer 2000, 2007.
105. Adolphs et al. 2001; Gigerenzer 2007; Jackson and Decety 2004; Schulkin 2007; Barton 2004.
106. Le Doux 1996; Swanson 2000; Amaral et al. 1992/2000; Aggleton 1992/2000; Emery 2000.
107. Barger, et al. 2007; Barton, et al. 2003.
108. Rolls 2000; Emery and Amaral 2000.
109. Zink et al. 2011; Schutz 1967; Schulkin 2004.
110. Carter et al. 1997/1999.
111. Savaskan et al. 2008.
112. Zak 2008.
113. Kosfeld et al. 2005; Petrovic et al. 2008; Domes et al. 2007; Singer et al. 2008.
114. Thorpe 1974; Marler 2000.
115. Darwin 1872/1998.
116. Gurney and Konishi 1980; Marler et al. 1988.
117. Marler 2000.
118. Helmholtz 1873; Von Bekesy 1970.

Chapter 3: Human Song: Dopamine, Syntax, and Morphology

1. E.g., Lashley 1951.
2. Sacks 2008.
3. Brown 2008.
4. Lashley 1951.
5. Dunbar 1996; Seyfarth et al. 2005.
6. Levinson and Jaisson 2006; Wittgenstein 1953/1968.
7. Corbalis 2002.
8. Goldin-Meadow 1999.
9. Schulkin 2007; Barton 2006.
10. Tallis 2010. See also Hatten 2004.
11. Corbalis 2002; Goldin-Meadow et al. 2009.
12. Broaders and Goldin-Meadow 2009.
13. Corballis 2002.
14. Peirce 1903–1912/1977; see also Fiske 1991; Tarasti 1994; Cumming 1999; Cumming 2000; Lidov 2005; Short 2007.
15. Clark 1998.
16. Morris 1938/1979; Lakoff and Johnson 1999; Eco 1976.

17. Peirce 1903–1912/1977.
18. Ullman 2001.
19. Diamond 2001.
20. Swanson 2000.
21. Marsden 1984.
22. E.g., Grahn 2009.
23. Swanson 2000.
24. E.g., Kelley 1999.
25. Kelley 2004; Schultz 2002.
26. McIntosh et al. 1997; Thaut et al. 2001.
27. Sterling 2004.
28. Kennedy 1964
29. Schultz 2002.
30. Kelley 1999.
31. Ullman et al. 1997 Graybiel et al. 1994.
32. Marsden 1984.
33. Lashley 1951.
34. Zatorre 2001.
35. Schultz 2002.
36. Tremblay et al. 1998.
37. Schultz 2002.
38. Fiorillo et al. 2003.
39. Juslin and Sloboda 2001.
40. Salimpoor et al. 2011.
41. Kelley 1999.
42. Smith et al. 2011.
43. Berridge and Robinson 1998.
44. Berns et al. 2010. See also Menon and Levitin 2005; Blood et al. 1999.
45. Kelley 1999.
46. Ibid.
47. Ullman et al. 1997; Pinker 1994.
48. Marler and Hamilton 1966.
49. Janata and Grafton 2003; Meyer 1967; Clarke 2008.
50. Janata 2005.
51. Rathelot and Strick 2009; Zatorre 2001.
52. Alheid and Heimer 1988; Kelley 2004; Nauta and Freitag 1986.
53. Swanson and Petrovich 1998.
54. Swanson 2000, 2003.
55. Graybiel 1998.
56. Swanson 2003.
57. E.g., Lashley 1951.
58. Berridge et al. 2005.
59. Aldridge and Berridge 1998.
60. Marsden and Obeso 1994; Schwartz et al. 2011.
61. Maess et al. 2001; Salimpoor et al. 2011.
62. Berns et al. 2001; Vuust and Kringelbach 2010.

63. O'Doherty et al. 2004.
64. Berridge 2006; Schultz 2002.
65. Salimpoor, et al. 2011.
66. Grahn 2009.
67. Ibid.; Berridge 2006.
68. Sacks 2008, p. 275.
69. Jackson 1884/1958.
70. Barton 2006; Schulkin2007; Jackson and Decety 2004.
71. Thaut and McIntosh 2010.
72. Thaut et al. 2001.
73. Patel et al. 2008.
74. Sacks 2008.
75. Thaut 2005; Koelsch 2006; Hillecke et al. 2005.
76. Grocke and Wigram 2007; Horgen 2000; Koelsch et al. 2010; Koelsch et al. 2011; Morrison, et al. 2008.
77. Huron 2008.
78. Lakoff and Johnson 1999.
79. Repp 2005.
80. Penune et al. 2005.
81. Rickard 2009; Repp 2005; Brown and Dissanayake 2006.
82. Grahn 2009.
83. Sacks 2008, p. 276.
84. E.g., Martin et al. 1996; Ullman 2004.
85. Squire 2004.
86. Ryle 1949/1990; Mishkin and Petri 1984; Squire et al. 1993.
87. Ullman 2001.
88. Ibid.
89. Broca 1863; Wernicke 1874; Pulvermuller 2002; Chomsky 1965; Pinker 1994.
90. Miranda and Ullman 2007.
91. Ullman 2001; Berridge et al. 2005.
92. Koelsch 2011.
93. Schon et al. 2005.
94. Pfordresher and Brown 2009.
95. Koelsch et al. 2002.
96. Pulvermuller 2002.
97. Maess et al. 2001.
98. Sluming et al. 2007.
99. Sluming et al. 2002.
100. Fadiga et al. 2009.
101. Martin et al. 1996; Ullman and Pierpont 2005.
102. Friederici 2002.
103. Ullman 2001.
104. Bharucha et al. 2006; Jackendoff and Lerdahl 2006; Sloboda 2000.
105. See Schultz 2002; Meyer 1973.
106. Friederici et al. 2000; Koelsch et al. 2004.

107. Gallagher and Holland 1994.
108. LeDoux 1996, 2000; Parkinson et al. 2001.
109. Schwartz et al. 2003; Beaton et al. 2010.
110. McGaugh et al. 1996; Young et al. 1999; LeDoux 1996; Gallagher and Holland 1994.
111. Koelsch et al. 2008.
112. Fried et al. 2001.
113. Krumhans 2002; Curtis and Bharucha 2009.
114. Pfordhresher and Brown 2009.
115. Daffner et al. 2000.
116. Swanson 2000.
117. Elliott et al. 2003.
118. Ruiz et al. 2009.
119. Ullman et al. 1997; Ullman 2001, 2004.
120. Maidhof et al. 2010.
121. Leaver et al. 2009.
122. Rosner and Meyer 1986; Cutting and Rosner 1974.
123. Dewey 1934; Meyer 1956; Schulkin 2007.
124. Rozin 1998.
125. Berns et al. 2001.

Chapter 4: Musical Expectations, Probability, and Aesthetics

1. Dewey 1929/1960, p. 169.
2. Meyer 1956.
3. Dewey 1934/1958, p. 236.
4. Sloboda 2005; Panksepp and Bernatzky 2002; Vuust and Kringelbach 2010.
5. Meyer 1965; Schulkin 2004; Perlovsky 2010.
6. Meyer 1856, p. 236.
7. Peirce 1898; Dewey 1938; Loewenstein 1994.
8. Gjerdingen 2007. See also his *Meyer and Music Usage* for which the above quote is derived. See also his other student, Narmour 2008.
9. Nietzsche 1871/1927.
10. Dewey 1929/1960, p. 168.
11. See Meyer 1956. See also notes on Meyer by D. Huron.
12. Huron 2008; Cross 2010.
13. Peirce 1898; Morris 1938/1979.
14. Dewey 1934; Meyer 1973.
15. Huron 2008.
16. Ibid., p. 270.
17. Rozin 1998.
18. Hacking 1999.
19. Meyer 1973.
20. Narmour 1990; Gjerdingen 1990; Russo and Cuddy 1999; Cuddy and Lunney 2005.
21. Thompson et al. 1997.

22. Vuust et al. 2009.
23. Meyer 1973; Huron 2009; Sloboda 1991; Gjerdingen 2007.
24. Dewey 1934/1958, p. 18.
25. Marler 2000; Geissman 2002; Brown et al. 2004.
26. Nettl 1948, 1956.
27. Miller 1959; Rescorla 1988.
28. Peirce 1878.
29. Dickinson 1980.
30. Rescorla and Wagner 1972.
31. Meyer 1967, p. 169.
32. Sterling 2004; Schulkin 2003.
33. Meyer, 1967, p. 353; see also Krumhansl 2002.
34. Sloboda 2008.
35. Berlyne 1960.
36. Tinbergen 1951.
37. Loewenstein 1994, 1996.
38. Freud 1915, p. 153, cited in Loewenstein 1996.
39. Dewey 1934/1958.
40. Piaget 1954; Hebb 1949.
41. Miller and Ogawa 1963.
42. Krieckhaus 1970.
43. Berylne 1960; Huron 2008.
44. James 1890/1952.
45. Dewey 1925/1989.
46. Berylne 1960.
47. Meyer 1973; Berylne 1960; Huron 2005; Sloboda 2000.
48. Loewenstein 1994.
49. James 1890/1952.
50. Dewey 1925/1989
51. Loewenstein 1994.
52. Miller 1957; Miller 1959.
53. Loewenstein, unpublished.
54. Loewenstein 1994.
55. Ibid.
56. Narmour 2008; Schulkin 2004; Gjerdingen 1999.
57. Peirce 1866, 1868; Meyer 1956.
58. Peirce 1998; Dewey 1929/1960.
59. Narmour 2008; Rozin 1976.
60. Rozin and Rozin 2006.
61. Narmour 2008.
62. Dewey 1894; Meyer, 1956; Lerdahl and Jackendoff 1983; Krumhansl 2002.
63. Loewenstein and Lerner 2003.
64. James 1890/1952; Hebb 1949; Loewenstein 1996.
65. Elster 2000, p. 205.
66. Davies 1994.

67. Sloboda 2000; Temperley 2001.
68. Temperley 2004, p. 3.
69. Temperley 2004; Sloboda 1991.
70. Narmour 1990.
71. See Sabini and Schulkin 1994.
72. Sloboda 2008, p. 2.
73. Sloboda 1991; Steinbeis et al. 2006; Rosner and Narmour 1992.
74. Meyer 1956; see Dewey 1894.
75. Sloboda 1991.
76. Juslin and Vastfall 2008; see also Zentner 2011.
77. Hammond and Stewart.
78. Juslin 2001.
79. Steinbeis et al. 2006.
80. Hebb 1949.
81. Lashley 1951.
82. Ibid., p. 506.
83. Dewey 1934/1958.
84. See Craig 1918.
85. See also MacDowell and Mandler 1989.
86. Dewey 1925/1989.
87. See also Meyer 1967, 1973.
88. Wittgenstein 1953/1968.
89. Cf. Langer 1957; Weber 1931; Lerdhal 1992.
90. Smith 1987; Smith and Witt 1989.
91. Smith and Melara 1990.
92. Kosslyn 1986; Farah 1984; Shepard 1994.
93. Rozin 1976.
94. Mandler 2004; Huron 2005; Vuust and Kringelbach 2010.
95. James 1890/1952; Cannon 1927.
96. Panksepp 1995.
97. Sloboda 1985/2000.
98. Sloboda 2000.
99. Simon 1982.
100. Gigerenzer 2000; Huron 2008.
101. Huron 2008, p. 94.
102. Falk 2000.
103. Friederici 2002; Patel et al. 2008; Patel et al. 1998; Raffman 1993; Lerdahl and Jackendoff 1999.
104. Patel 2003; Friederici 2002.
105. Patel 2005.
106. E.g., Broca's region; Patel et al. 1998; Patel 2003; Freiderici 2002.
107. Cook 1998; Patel 2003.
108. Sergent et al. 1992; Koelsch et al. 2000; Koelsch et al. 2002; Koelsch 2006; Maess et al. 2001; Gray et al. 2001.
109. E.g., Grahn and Rowe 2009; Koelsch 2006.

110. Koelsch 2006; Bengtsson et al. 2009; Brown and Martinez 2006.
111. Maess et al. 2001.
112. Koelsch 2011.
113. Patel et al. 1998.
114. Koelsch et al. (2002); Koelsch et al. 2003; Patel 2003.
115. Peirce 1899/1992; Brunswik 1943; Gigerenzer 2000; Kahneman et al. 1982.
116. Goodman 1955/1978.
117. Dewey 1934/1985; Goodman 1968.
118. Meyer 1956; Huron 2008; Smith and Melara 1990; Juslin 2001.
119. Hacking 1964.
120. Gigerenzer 1991; Baron 1988/2008.
121. Peirce 1898.
122. Gjerdingen 1999; Clarke, E. 2008.
123. Repp 2005.
124. Gibson 1966; Gigerenzer 2000.
125. Goldstein and Gigerenzer.
126. Clarke, E. 2008.
127. Clark, A. 1998.
128. Dehaene 1997.
129. Leslie et al. 2008; Carey 2009; Peirce 1899/1992, 2000, 2009.
130. Lakoff and Numez 2000; Dehaene 1997.
131. Friederici 2002.
132. Lakoffand Numez 2000.
133. Whitehead 1919/1982.
134. Kline 1959.
135. Lakoff and Numez 2000.
136. Clark, A. 1998.
137. Dehaene et al. 2006.
138. Desmurget et al. 2009.
139. Knops et al. 2009.
140. Brunswik 1943; Gigerenzer 2000.
141. Dewey 1934/1958; Meyer 1956, 1967.
142. Wallin et al. 2000.
143. Smith 1997; Smith, D. J. et al. 1994.
144. Kant 1792, 1951.
145. Clark, A. 1998; Noe 2004; Wheeler 1995; Donald 2001.
146. Nussbaum 2001.
147. Langer 1973.
148. Koelsch et al. 2000.
149. Schmidt and Trainor 2001.
150. Blood et al. 1999.
151. Davidson et al. 2000.
152. See also Cardinal et al. 2002, for roughly the same suggestion for the origins of the emotions.

Chapter 5: Musical Expression, Memory, and the Brain

1. Kant 1792/1951.
2. Lieberman 1984; Cross 2009a and b.
3. Darwin 1871/1874.
4. Darwin 1859/1958.
5. James 1890/1952, p. 389.
6. James 1890/1952, Vol. 2, p. 39.
7. Palmer et al. 2001.
8. Gallistel 1980.
9. Langer 1973, p. 45.
10. Tinbergen 1951.
11. Cf. Pinker and Jackendoff 2005; Fitch 2005.
12. Pinker 1994.
13. Cook 1998.
14. Blacking 1973.
15. Humbolt 1836/1971; Chomsky 1965.
16. Diderot 1755/1964; Darwin 1872/1998; Spencer 1855, 1852 see Cross 2009.
17. Fitch 2005.
18. Tinbergen 1951.
19. Garcia, et al. 1974; Rozin 1976.
20. Gigerenzer 2000; Schulkin 2009.
21. Thorpe 1974.
22. Geissmann 2000.
23. Panksepp 1995.
24. Dewey 1925/1989.
25. See Meyer 1956; Dewey 1934/1958; Mead 1934.
26. Fredierici 2002.
27. Galison 1988; Heelan and Schulkin 1998.
28. Dewey 1934/1958; Lakoff and Johnson 2000; Schulkin 2009.
29. Dewey 1934/1958.
30. Ibid., p. 14.
31. James 1890, Vol. 1, p. 451.
32. Meyer 1956.
33. Ibid., p. 93.
34. James 1890/1952; Tulving and Craik 2000.
35. Janata 2005.
36. Mandler 2004; Palmer, et al. 2009.
37. Schacter and Tulving 1994.
38. E.g., Peirce 1899/1992; Clark, A. 1997.
39. Gibson 1966, 1979.
40. Clarke, E. 2008.
41. See Carey 1985/1987; Kornblith 1993.
42. E.g., Gibson 1979; Clark 1997.
43. Clark, A. 1997; Heelan and Schulkin 1998.
44. Feder 2004, p. 19; Fischer 2011.

45. James 1890/1952; Schacter and Tulving 1994.
46. Zattore 2001; Eichenbaum and Cohen 2001.
47. Koelsch et al. 2002; Janata 2005.
48. Janata 2005.
49. Schneider et al. 2002; Zattore 2001.
50. Jackson 1884/1958; Critchley and Critchley 1998; Sacks 2008.
51. Sacks 2008, p. 21.
52. Ibid., p. 27.
53. Ibid., p. 28.
54. Wynn and Coolidge 2008.
55. McGaugh 2003.
56. Pashler 1998.
57. Desimone 1996.
58. Kraus and Chandrasekaran 2010.
59. James 1890, Vol. 1, p. 434.
60. Gjerdingen 1999.
61. Rozin 1976; Squire 1987; Tulving and Craik 2000.
62. E.g., Squire and Zola 1996.
63. E.g., Rolls, Treves, and Tovee, 1997.
64. Martin 1998; Jeannerod 1999; Rizzolatti and Arbib 1998.
65. Martin 2007.
66. E.g., Perrett and Emery 1994.
67. Clarke, E. 2008.
68. Palmer et al. 2009.
69. Martin et al. 1996.
70. Zatorre 2002.
71. Zbikowski 2010; Clarke, E. 2008; Johnson 2007.
72. Bachorik et al. 2009.
73. Gallese et al. 1996.
74. Frith 2007; Meyer 1973. C. S. Peirce quote cited in Kruse 2005.
75. Rizzolatti and Luppino 2001; Decety et al. 1994; Cox 2011.
76. E.g., Martin et al. 1995, 1996; Chao et al. 1999; see also Johnson 2007.
77. Pfordresher 2003.
78. Rizzolatti and Arbib 1998.
79. Baumann et al. 2005.
80. Peacocke 2009; Chapados and Levitin 2008.
81. E.g., Ray Charles.
82. Premack 1990.
83. Chen et al. 2008.
84. Stewart et al. 2003.
85. Ozdemir et al. 2006; Besson et al. 1998; Abrams et al. 2011.
86. Wong et al. 2007; Lee et al. 2009; Satoh et al. 2001. Cohen et al. 2011.
87. Penhune 2011; Penhune et al. 2005; Marin 2009; Stegemoller et al. 2008.
88. Kraus and Chandrasekaran 2010; Tramo et al. 2002; Lee et al. 2009; Musacchia et al. 2007.

89. Gaser and Schlaug 2003.
90. Schlaug et al. 1995.
91. Loui et al. 2010.
92. Zattore et al. 2002.
93. Peretz et al. 2009.
94. Ibid.
95. Zattore et al. 2005; Schneider et al. 2002; Gaser and Schlaug 2003; Keenan et al. 2001.
96. Koelsch, 2011; Zattore 2001.
97. Zattore 2001.
98. Dewey 1894; Meyer 1973.
99. Koelsch 2011, 2005.
100. Demorest et al. 2010.
101. Schellenberg et al. 2002; Curtis and Bharucha 2009.
102. Pfordresher 2003; Cohen et al. 1989.
103. Nettl 1956.
104. James 1890; Dennet 1987.
105. Cf., Spencer 1855; Darwin 1872/1998; Diderot 1755/1964; also see Cross 2009; Koelsch 2011; Kant 1792/1951.
106. Cross 2009.
107. Dewey 1925/1989; Merleau-Ponty 1942/1967; Lakoff and Johnson 1999.

Chapter 6: Development, Music, and Social Contact

1. Zentner and Eerola 2010.
2. Hannon and Trainor 2007.
3. Trainor and Trehub 1994; Trainor et al. 2002.
4. Carey 1985; Kagan 1984; Keil 1979; Premack 1990.
5. Krumhansl 2002; Gosselin et al. 2006; Blood et al. 1999.
6. Perry et al. 1987; Grossmann et al. 2010.
7. Cf., Helmholtz 1873; Schoenberg 1975; see also Daynes 2010; Reich 1974.
8. Piaget 1954; Carey 1985/1987; Kagan 1984; Premack and Premack 1995.
9. Lahav et al. 2005; Merker et al. 2009.
10. Hannon and Trainor 2007.
11. Zetner and Kagan 1996; Hannon and Trehub 2005; Trainor et al. 2002.
12. Kagan 2002; Falk 2000; Schmidt and Trainor 2001.
13. Hannon and Trehub 2005; Perani et al. 2010; Winkler et al. 2009.
14. Jackendoff and Lerdahl 2006.
15. Zentner and Kagan 1996.
16. Peretz 2006.
17. Stewart 2006; McDonald and Stewart 2008; McDonald 2006.
18. Peretz 2006.
19. Sergent 1993; Jourdain 2002.
20. Basso and Capitani 1985.
21. Cuddy et al. 2005.

22. Stewart et al. 2006; McDonald 2004.
23. Peretz 2006.
24. Hyde et al. 2007.
25. Landau and Hoffman 2005.
26. Reiss et al. 2005.
27. Baron-Cohen 1995/2000; Porges 2004.
28. Heaton 2009.
29. Hollinger et al. 2003; Wan et al. 2010.
30. Malloch and Trevarthen 2009; Schulkin 2004.
31. Galaburda et al. 2001.
32. Reiss et al. 2005; Galaburda et al. 2001.
33. Haas et al. 2009.
34. Don et al. 1999; Levitin et al. 2004.
35. Levitin et al. 2004; Bhatara et al. 2010.
36. Huron 2001; Levitin et al. 2004.
37. Levitin and Bellugi 2006.
38. Levitin et al. 2003.
39. Thornton-Wells et al. 1997.
40. Dykins et al. 2005.
41. Gibson 1979; Garcia et al. 1974; Rozin 1976; Rosati et al. 2007.
42. Premack 1990; Dennett 1987.
43. Dennett 1987; Premack 1990; Tomasello et al. 1993; Tomasello et al. 2004.
44. Tomasello 1999; Tomasello et al. 2004.
45. Kagan 1984; Tomasello et al. 1993; Tomasello 1999; Tomasello et al. 2004; Baron-Cohen et al. 1993/2000.
46. E.g., Mead 1934; Vygotsky 1934/1979; Humphrey 1976, 1992.
47. E.g., Mead 1934.
48. Vygotsky 1934/1979.
49. Humphrey 1976, 1992.
50. Falk 2005.
51. Insel 2010.
52. Rozin 1976, 1998.
53. Sterelny 2007.
54. Power and Schulkin 2009.
55. Jackson 1884/1958; James 1890/1952; Rozin 1976.
56. Thorpe et al. 2007; Dunbar 2003; Corbaliss and Lea 1999.
57. E.g., Silk 2007.
58. Ibid.; O'Connor 2007.
59. E.g., Gould and Elderidge 1977; Gould 2002; Wood 2000.
60. Rakic 1988, 2002.
61. Walker 1962/1982.
62. Whiten and Van Schaik 2007.
63. Barton 2006; Dunbar 1996, 1998, 2003; Dunbar and Shultz 2007.
64. Cheney and Seyfarth 1990, 2007; Dunbar and Shultz 2007; Barton 2004, 2006; Byrne and Corp 2004.
65. Kirschner and Tomasello 2010, 2009.
66. Byrne 1995; Byrne and Whiten 1988.

67. Reader and Laland 2002; Barton2004; Schulkin 2007.
68. Gallison 1988; Lakoff and Johnson1999); Heelan and Schulkin 1998.
69. Reader and Laland 2002.
70. Byrne and Corp 2004.
71. Ungerleider and Mishkin 1982; Goodale and Milner 1992.
72. Rauschecker and Scott 2009.
73. Passingham 2008; Kakei et al. 2001. See Wise 1985. Also see Jackson 1884/1958; Zattore et al. 2002, 2007; Rauschecker and Scott 2009.
74. Rolls and Treves 1998; Emery 2000.
75. Livingstone and Thompson 2009.
76. Galison 1988.
77. E.g., Fisher; see Jablonka and Lamb 1995.
78. Dobzhansky 1962; see Holliday 2006; Crews 2011.
79. Darwin 1859/1958; James 1890/1952.
80. Holliday and Pugh 1998; Holliday 2002.
81. Holliday and Ho 1998; Keverne and Curley 2008.
82. Keverne and Curley 2008.
83. Weaver et al. 2004.
84. Carter et al. 1997/1999.
85. Hillecke et al. 2005.
86. Weaver et al. 2004.
87. Auger et al. 2011.
88. Mead 1934.
89. Sacks 2010.
90. Ibid., p. 31.

Chapter 7: Music and Dance

1. Langer 1957.
2. Langer 1953, p. 204; see also Schultz 1964; Bellah 2011.
3. Langer 1953, p. 80.
4. Levi 1967; Mead1928/1972.
5. E.g., Burgess and Haynes 2004.
6. Dipert 1983; see also the following pragmatist articles undercutting the cognitive and non/cognitive distinction: Parrott and Schulkin 1993; Kruse 2007; Langer, 1957; and Dewey 1934.
7. Sachs 1937/1963.
8. Von Frisch 1953.
9. Richter 1965/1979; Moore 1992.
10. Richter 1965/1979.
11. Thaut 2003.
12. Richter 1965/1979.
13. Buzaki 2006.
14. Simmons and Martin 2011.
15. James 1993; Thomas 2001; Cook 1998; Guthrie 1955.
16. Mithen 1996.
17. Peirce 1898; Hanson 1971; Hanson 1958/1972.

18. Rozin 1976.
19. Ibid., 1998; Dunbar 1998.
20. Cross 2009; Huron 2009.
21. Ramachandran and Hirstein 1999; Zeki et al. 1993.
22. Mithen 1996.
23. Kant 1792.
24. Ibid., p. 108.
25. See Johnson 1987; Bellah 2011; Schulkin 2004.
26. Langer 1957, p. 80.
27. Mithen 1996; Budd 2002. Also see Croce's masterful history of aesthetics in the classical and premodern period: Croce 1909/1983.
28. Mithen 1996; Humphrey 1973.
29. Mithen 1996; Rozin 1976, 1998.
30. Premack 1982; personal communication.
31. Humphrey 1959; Delza 1996; Devi 1962.
32. Sachs 1937/1963.
33. Von Frisch 1953; Brown, S. 2003.
34. Sachs 1937/1963, p. 83.
35. Clarke, E. 2008; Lloyd 1949; Hanslick 1854.
36. Wiora 1965; Sachs 1937/1963; Nettl 1956, 2005.
37. Scheijen 2009; Taruskin 1991; Joseph 2011.
38. See Nichols 2011.
39. Morgan 1941/1980.
40. Copland 1937/1957. See also Copland 2006.
41. Siegel 1979; Lloyd 1974.
42. Arsemault 2009.
43. E.g., Varela et al. 1991; Lakoff and Johnson 1999; Schulkin 2004; Gallagher 2005.
44. Zbikowski 2011a and b.
45. Johnson 1993.
46. See also James 1890/1952; Dewey 1896; Schulkin 2009.
47. Wilson 2002; Barsalou 2003; Clark, A. 1998, 1999; Noe 2004.
48. See also, e.g., Dewey 1896; Barton 2004; Schulkin 2007; Jeannerod 1997; Berthoz 2002.
49. Johnson 2007.
50. Johnson and Larson 2003.
51. Ibid.
52. Meyer 1967
53. Tomasselo and Carpenter 2007; Gallagher 2005.
54. E.g., James 1890/1952; Merlau-Ponty 1962; Damasio 1994; Schulkin 2004; Berthoz 2000.
55. Cross 2009; Miell et al. 2005.
56. Clarke, E. 2008.
57. Ibid., p. 63.
58. Gigerenzer 2000; Schulkin 2009; Sevdalis and Keller 2011.
59. Lakoff and Johnson 1999; Zbikowski 2010.
60. See Hutchinson 1977.

61. Zbikowski 2010.
62. Ibid.
63. Zbikowski 2008.
64. Epstein 2008, p. 153.
65. Suskin 2010; Bordman 1992.
66. Gallese 2007.
67. Brown and Parsons 2008.
68. Atran 1990/1996; Boyer 1990.
69. Atran 1990/1996.
70. Keil 1989; Carey 2009; Schulkin 2009.
71. Richardson 2010.
72. Kant 1787/1965.
73. Cooke 1959/1964; Trimble 2007.
74. T. S. Eliot 1943.
75. Lakoff and Johnson 1999.
76. Humphrey 1959, p. 104.
77. Atran 1990/1996; Sachs 1937/1963; Nettl 2005; Mead 1928/1972; Keltner and Haidt 2003.
78. Meyer 1967, p. 170.
79. Spitzer 2004; Levi-Strauss 1969; Lakoff and Johnson 1999.
80. Bordman 1992; Suskin 2010.

Conclusion: Music and Well-Being

1. Cross 1999; Morley 2003.
2. Brown 2007.
3. Dewey 1934. See also Sloboda 1985/2000; Huron 2008; Temperley 2001.
4. Neuhaus et al. 2006; Rosen 2010. See also Dewey 1934.
5. Lambert 1997.
6. Clarke, E. 2008.
7. Lashley 1951; Janata 2005; Patel 2008; Zatorre 2001.
8. Gibson 1966; Clarke, E. 2008.
9. Koelsch 2011; see also Levman 1992.
10. Jentschke et al. 2008.
11. Cross 2010.
12. Whitehead 1929/1958; Rousseau 1966.
13. Huron 2005; Merker 2005.
14. Han, et al. 2011; Falk 2000 ; Patel 2010; Koelsch 2011; Zatorre et al. 2002.
15. Chapin et al. 2010.
16. Huron 2008; see also Sloboda 1991.
17. Meyer 1956, p. 261.
18. Peirce 1898; Dewey 1934/1958; Kruse 2005.
19. Dewey 1934/1958, p. 236.
20. Mithen 1996.
21. Fromm 1973.
22. Sacks 2008.

23. Nietzsche 1871/1927, p. 167.
24. See Liebert 2004.
25. Schopenhauer 1844/1966, p. 448.
26. Rumph 2011.
27. Wagner 1911.
28. Kitcher and Schacht 2004.
29. Nietzsche 1871/1927.
30. Frederickson 2004.
31. Steinbeis and Koelsch 2007.
32. Engel and Keller 2011; jazz musicans who improvise in peformance have greater activation of the amygdala.

References

http://www.musicbrain.com/people.php#publications

Abrams, D. A., Bhatara, A., Ryah, S., Balaban, E., Levitin, D., and Menon, V. (2011). Decoding temporal structure in music and speech relies on shared brain resources but elicits different fine-scale spatial patterns. *Cerebral Cortex, 21*: 1507–1518.

Adler, D. S. (2009). The earliest musical tradition. *Nature, 460*: 695–696.

Adler, D. S., Bar-Yosef, O., Belfer-Cohen, A., Tushabramishvili, N., Boaretto, E., Mercier, N., et al. (2008). Dating the demise: Neanderthal extinction and the establishment of modern humans in the southern Caucasus. *Journal of Human Evolution, 55*: 817–833.

Adolphs, R. (1999). Social cognition and the human brain. *Trends in Cognitive Sciences, 3*: 469–479.

Adolphs, R., Denburg, N. L., and Tranel, D. (2001). The amygdala's role in long-term declarative memory for gist and detail. *Behavioral Neuroscience, 115*: 983–992.

Adorno, T. (1976). *Introduction to the Sociology of Music.* New York: Seabury Press.

Aggleton, J. (1992, 2000). *The Amygdala.* Oxford: Oxford University Press.

Albers, H. E., Hennessey, A. C., and Whitman, C. D. (1992). Vasopressin and the regulation of hamster social behavior. *Annals New York Academy of Sciences, 652*: 227–242.

Aldridge, J. W. and Berridge, K. C. (1998). Coding of serial order by neostriatal neurons: a 'natural action' approach to movement sequence. *Journal of Neuroscience, 18*: 2777–2787.

Aldridge, J. W. and Berridge, K. C. (2010). Neural coding of pleasure: "rose-tinted glasses" of the ventral palladium. In M. L. Kringelbach and K. C. Berridge (eds.), *Pleasures of the Brain* (pp. 62–73). New York: Oxford University Press.

Aldridge, J. W., Berridge, K. C., and Rosen, A. R. (2004). Basal ganglia neural mechanisms of natural movement sequences. *Canadian Journal of Physiological Pharmacology, 82*: 732–739.

Alembert, J. (1751/1963). *Preliminary Discourse to the Encyclopedia of Diderot.* New York: Library of Liberal Arts.

Alheid, G. F. and Heimer, L. (1988). New perspectives in basal forebrain organization of special relevance for neuropsychiatric disorders: The striatopallidal, amygdaloid and corticopetal components of substantia innominata. *Neuroscience, 27*: 1–39.

Allen, J. S. (2009). *Human Evolution and the Organ of Mind*. Cambridge, MA: Harvard University Press.

Altenmuller, E. O. (2001). How many music centers are in the brain? *Annals of the New York Academy of Sciences, 930*: 273–280.

Altman, J. (1966). Autoradiographic and histological studies of postnatal neurogenesis. *Journal of Comparative Neurology, 124*: 431–474.

Alvarez-Buylla, A., Theelen, M., and Nottebohm, F. (1988). Birth of projection neurons in the higher vocal center of the canary forebrain before, during, and after song learning. *Proceedings of the National Academy of Sciences Neurobiology, 85*: 8722–8726.

Amaral, D. G., Price, J. L., Pitkanen, A., and Carmichael, S. T. (1992, 2000). Anatomical organization of the primate amygdaloid complex. In J. P. Aggleton (ed.), *The Amygdala: Neurobiological Aspects of Emotion, Memory and Mental Dysfunction* (pp. 1–66). New York: Wiley-Liss.

Aristotle (1968). *De Anima* (translated by D. Hamylin). Oxford: Oxford University Press.

Armstrong, E. A. (1965). *Bird Display and Behavior*. New York: Dover Press.

Arnheim, R. (1974). *Art and Visual Perception*. Berkeley and Los Angeles: University of California Press.

Arnold, A. P. (2002). Concepts of genetic and hormonal induction of vertebrate sexual differentiation in the twentieth century with special reference to the brain. In D. W. Pfaff, et al. (eds.), *Hormones, Brain and Behavior* (pp. 105–135). New York: Elsevier Press.

Arsemault, R. (2009). *The Sound of Freedom*. New York: Bloomsbury Press.

Asfar, B., White, T., Lovejoy, O., Latimer, B., Simpson, S., and Suwa, G. (1999). *Autralopithecus garhi*: A new species of early hominid from Ethiopia. *Science, 284*: 629–634.

Atran, S. (1990, 1996). *Cognitive Foundations of Natural History*. New York: Cambridge University Press.

Atran, S., Medin, D. L., and Ross, N. O. (2005). The cultural mind. *Psychological Review, 112*: 744–766.

Auger, C. K., Coss, D., Auger, A. P., and Forbes-Lorman, R. M. (2011). Epigenetic control of vasopressin expression is maintained by steroid hormones in the adult male rat brain. *Proceedings of the National Academy of Sciences of the United States, 108*: 4242–4247.

Babbitt, I. (1955). *Rousseau and Romanticism*. New York: Meridian Books.

Bachorik, J. P., Bangert, M., Loui, P., Larke, J., Berger, J., Rowe, R., et al. (2009). Emotion in motion: Investigating the time-course of emotional judgments of musical stimuli. *Music Perception, 26*: 355–364.

Ball, G. F., and Balthazart, J. (2002). Neuroendocrine mechanisms regulating reproductive cycles and reproductive behavior in birds. In D. W. Pfaff, et al. (eds.), *Hormones, Brain and Behavior* (pp. 649–798). New York: Academic Press.

Balter, M. (2010). Animal communication helps reveal roots of language. *Science, 328*: 696–971.

Barger, N., Stefanacci, L., and Semendeferi, K. (2007). A comparative volumetric analysis of the amygdaloid complex and basolateral division in the human and ape brain. *American Journal of Physical Anthropology, 134*: 392–403.

Barham, L. and Mitchell, P. (2008). *The First Africans*. Cambridge: Cambridge University Press.
Baron, J. (1988, 2008). *Thinking and Deciding*. Cambridge: Cambridge University Press.
Baron-Cohen, S. (1995, 2000). *Mindblindness*. Cambridge, MA: MIT Press.
Baron-Cohen, S., Tager-Flushberg, H., and Cohen, D. J. (1993, 2000). *Understanding Other Minds*. Oxford: Oxford University Press.
Barrett, L. and Henzi, P. (2005). The social nature of primate cognition. *Proceedings of the Royal Society B: Biological Sciences, 272*: 1865–1875.
Barrett, L., Henzi, P., and Rendall, D. (2007). Social brains, simple minds: does social complexity really require cognitive complexity? *Philosophical Transactions of the Royal Society of London B: Biological Sciences, 362*: 561–575.
Barsalou, L. W. (2003). Abstraction in perceptual symbol systems. *Philosophical Transactions of the Royal Society of London B: Biological Sciences, 358*: 1177–1187.
Bartlett, J. C. and Dowling, J. W. (1988). Scale structure and similarity of melodies. *Music Perception, 5*: 285–314.
Bartok, B. (1976). *Essays*. London: St. Martin's Press.
Barton, R. A. (2004). Binocularity and brain evolution in primates. *Proceedings of the National Academy of Sciences of the United States, 101*: 10113–10115.
Barton, R. A. (2006). Primate brain evolution: Integrating comparative neurophysiological and ethological data. *Evolutionary Anthropology, 15*: 224–236.
Barton, R. A., Aggleton, J. P., and Grenyer, R. (2003). Evolutionary coherence of the mammalian amygdala. *Proceedings of the Royal Society of London B: Biological Sciences, 270*: 539–543.
Bartz, J. A., Zaki, J., Ochsner, K. N., Bolger, N., Kolevzon, A., Ludwig, N., et al. (2010). Effects of oxytocin on recollections of maternal care and closeness. *Proceedings of the National Academy of Sciences of the United States, 107*: 21371–21375.
Barzun, J. (1949). *Berloiz and the Romantic Century*. New York: Columbia University Press.
Basso, E. (1985). *A Musical View of the Universe*. Philadelphia: University of Pennsylvania Press.
Basso, E. and Capitani, E. (1985). Spared musical abilities in a conductor with global aphasia and ideomotor apraxia. *Journal of Neurology, Neurosurgery and Psychiatry, 48*: 407–412.
Bauman, S., Koeneke, S., Meyer, M., and Jancke, L. (2005). A network for sensory-motor integration. *Annals of the New York Academy of Sciences, 1060*: 186–188.
Beaton, E. A., Schmidt, L. A., Schulkin, J., and Hall, G. B. (2010). Neural correlates of implicit processing of facial emotions in shy adults. *Personality and Individual Differences, 49*: 755–761.
Békésy, G. von (1959). *Experiments in Hearing*. New York: McGraw-Hill.
Bellah, R. N. (2011). *Religion in Human Evolution*. Cambridge, MA: Harvard University Press.
Bengtsson, S. L., Nagy, Z., Skare, S., Forsman, L., Forssberg, H., and Ullen, F.

(2005). Extensive piano practicing has regionally specific effects on white matter development. *Nature Neuroscience, 8*: 1148–1150.

Bengtsson, S. L., Ullen, F., Ehrsson, H. H., Hasminoto, T., Kito, T., Naito, E., et al. (2009). Listening to rhythms activates motor and premotor cortices. *Cortex, 45*: 62–71.

Bentley, P. J. (1982). *Comparative Vertebrate Endocrinology*. Cambridge: Cambridge University Press.

Berlyne, D. E. (1954). A theory of human curiosity. *British Journal of Psychology, 45*(3): 180.

Berlyne, D. E. (1958). The influence of complexity and novelty in visual figures on orienting responses. *Journal of Experimental Psychology, 55*, 289-296.

Berlyne, D. E. (1966). Curiosity and Exploration. Downloaded at www.sciencemag.org.

Berlyne, D. E. (1970). Novelty, complexity, and hedonic value. *Perception & Psychophysics, 8*: 279–286.

Bernard, C. (1865, 1957). *An Introduction to the Study of Experimental Medicine*. New York: Dover Publications.

Berns, G. S. and Moore, S. E. (2011). A neural predictor of cultural popularity. *Journal of Consumer Psychology, 22*: 154–160.

Berns, G. S., Capra, C. M., Moore, S., and Noussair, C. (2010). Neural mechanisms of the influence of popularity on adolescent ratings of music. *NeuroImage, 49*: 2687–2696.

Berns, G. S., McClure, S. M., Pagnoni, G., and Montague, P. R. (2001). Predictability modulates human brain response to reward. *Journal of Neuroscience, 21*: 2793–2798.

Berridge, K. C. (2004). Motivation concepts in behavioral neuroscience. *Physiology & Behavior, 81*(2): 179–209.

Berridge, K. C. (2006). The debate over dopamine's role in reward: the case for incentive salience. *Psychopharmacology, 191*: 391–431.

Berridge, K. C., Aldridge, W. J., Houchard, K. R., and Zhuang, X. (2005). Sequential super-stereotypy of an instinctive fixed action pattern in hyper-dopaminergic mutant mice: a model of obsessive compulsive disorder and Tourette's. *BioMed Central Biology, 3*: 4.

Berridge, K. C. and Robinson, T. E. (1998). What is the role of dopamine in reward: Hedonic impact, reward learning or incentive salience? *Brain Research Reviews, 18*: 309–369.

Berthoz, A. (2002). *The Brain's Sense of Movement*. Cambridge, MA: Harvard University Press.

Berylne, D. E. (1960). *Conflict, Arousal and Curiosity*. New York: McGraw-Hill.

Besson, M., Faita, F., Peretz, I., Bonnel, A. M., and Requin, J. (1998) Singing in the brain. *Psychological Science, 9*: 494–498.

Bharucha, J. J., Curtis, M., and Paroo, K. (2006). Varieties of musical experience. *Cognition, 100*: 131–172.

Bhatara, A., Quintin, E. M., Levy, B., Bellugi, U., Fombone, E., and Levitin, D. J. (2010). Perception of emotion in musical performance in adolescents with autism spectrum disorders. *Autism Research, 2*: 214–225.

Blacking, J. (1973). *How Musical Is Man?* Seattle: University of Washington Press.

Blacking, J. (1981). Making artistic popular music: The goal of true folk. *Popular Music, 1*: 9–14.

Blake, E. and Cross, I. (2008). Flint tools as portable sound-producing objects in the upper Paleolithic context: An experimental study. *Experiencing Archaeology by Experiment*: 1–19.

Blakemore, S. and Decety, J. (2001). From the perception of action to the understanding of intention. *Neuroscience*, *2*: 561–567.

Blanning, T. (2008). *The Triumph of Music*. Cambridge, MA: Harvard University Press.

Bloch, E. (1974, 1985). *Essays in the Philosophy of Music*. Cambridge: Cambridge University Press.

Blood, A. J., Zatorre, R. J., Bermudez, P., and Evans, A. C. (1999). Emotional responses to pleasant and unpleasant music correlate with activity in paralimbic brain regions. *Nature Neuroscience*, *2*: 382–387.

Bolhuis, J. J. and Gahr, M. (2006). Neural mechanisms of birdsong memory. *Nature Reviews Neuroscience, 7*: 347–357.

Bordman, G. (1992). *American Musical Theatre*. Oxford: Oxford University Press.

Bowlby, J. (1988). *A Secure Base*. New York: Basic Books

Boyd, R. and Richerson, P. (2005). *Not By Genes Alone: How Culture Transformed Human Evolution*. Chicago: University of Chicago Press.

Boyer, P. (1990). *Tradition as Truth and Communication*. Cambridge: Cambridge University Press.

Brenowitz, E. A. (1991). Altered perception of species-specific song by female birds after lesions of a forebrain nucleus. *Science, 251*: 303–305.

Briggs, A. W., Good, J. M., Green, R. E., Krause, J., Maricic, T., Stenzel, U., et al. (2009). Targeted retrieval and analysis of five Neanderthal mtDNA genomes. *Science, 325*: 318–321.

Broad, K. D., Curley, J. P., and Keverne, E. B. (2006). Mother-infant bonding and the evolution of mammalian social relationships. *Philosophical Transactions of the Royal Society of London B: Biological Sciences*, *361*: 2199–2214.

Broaders, S. C. and Goldin-Meadow, S. (2009). Truth is at hand: How gesture adds information during investigative interviews. *Psychological Science, 2195*: 623–628.

Broca, P. (1863). Localization des functions cerebrales. Siege du langage articule. *Bulletins de la Societe d'Anthropologie (Paris), 4*: 200–203.

Brower, C. (2000). A cognitive theory of musical meaning. *Journal of Music Theory*, *44*: 323–379.

Brown, P. and Marsden, C. D. (1998). What do the basal ganglia do? *Lancet*, *351*: 1801–1804.

Brown, S. (2003). Biomusicology, and three biological paradoxes about music. *Bulletin of Psychology and the Arts*, *4*: 15–28.

Brown, S. (2006). The perpetual music track: The phenomenon of constant musical imagery. *Journal of Consciousness Studies, 13*: 25–44.

Brown, S. (2007). Contagious heterophony: A new theory about the origins of music. *Musicae Sceientiae, XI*: 3–26.

Brown, S. (2008). Music of language or language of music [Review of the *Music, Language and the Brain*]. *Trends in Cognitive Sciences, 12*: 246–247.

Brown, S. (2009). Review of *Music, Language and the Brain* by A. Patel. *Musicae Scientiae, 8*: 163–182.

Brown, S. and Dissanayake, E. (2006). Rituals and Ritualization. In S. Brown and U. Volgsten, (eds.), *Music and Manipulation: On the Social Uses and Social Control of Music* (pp. 31–56). Oxford: Berghahn Books.

Brown, S., Martinez, M. J., Hodges, D. A., Fox, P. T., and Parsons, L. M. (2004). The song system of the human brain. *Cognitive Brain Research, 20*: 363–375.

Brown, S., Martinez, M. J., and Parsons, L. M. (2006). The neural basis of human dance. *Cerebral Cortex, 16*: 1157–1167.

Brown, S., Ngan, E., and Liotti, M. (2007). A larynx area in the human motor cortex. *Cerebral Cortex, 18*: 837–845.

Brown, S. and Parsons, L. M. (2008). The neuroscience of dance. *Scientific American, 299*: 78–83.

Brunswik, E. (1943). Organismic achievement and environmental probability. *Psychological Review, 50*: 255–272.

Budd, M. (2002). *The Aesthetic Appreciation of Nature*. Oxford: Oxford University Press.

Burgess, G. and Haynes, B. (2004). *The Oboe*. New Haven, CT: Yale University Press.

Buzsaki, G. (2006). *Rhythms of the Brain*. Oxford: Oxford University Press.

Byrne, R. W. (1995). *The Thinking Ape: Evolutionary Origins of Intelligence*. Oxford: Oxford University Press.

Byrne, R. W. and Corp, N. (2004). Neocortex size predicts deception rate in primates. *Proceedings of the Royal Society of London, 271*: 1693–1699.

Byrne, R. W. and Whiten, A. (1988). *Machiavellian Intelligence: Social Expertise and the Evolution of Intellect in Monkeys, Apes and Humans*. Oxford: Oxford University Press.

Cage, J. (1961). *Silence: Lectures and Writings*. Middletown, CT: Wesleyan University Press.

Calvert, G. A., Bullmore, E. T., Brammer, M. J., Campbell, R., Williams, S. C. R., McQuire, P. K., et al. (1997). Activation of auditory cortex during silent lipreading. *Science, 276*: 593–596.

Cannon, W. B. (1927). The James-Lange theory of emotions: A critical examination and an alternative theory. *American Journal of Psychology, 39*: 106–124.

Cardinal, R. N., Parkinson, J. A., Hall, J., and Everitt, B. J. (2002). Emotion and motivation: The role of the amygdala, ventral striatum and prefrontal cortex. *Neuroscience and Biobehavioral Reviews, 26*: 321–352.

Carey, S. (1985, 1987). *Conceptual Change in Childhood*. Cambridge, MA: MIT Press.

Carey, S. (2009). *On the Origins of Concepts*. Oxford: Oxford University Press.

Carter, C. S. (2007). Sex differences in oxytocin and vasopressin: Implications for autism spectrum disorders? *Behavioural Brain Research, 176*: 170–186.

Carter, C. S. and Getz, L. L. (1993). Monogamy and the prairie vole. *Scientific American, 268*: 100–106.
Carter, S. C., Lederhendler, I. I., and Kirkpatrick, B. (eds.). (1997, 1999). *The Integrative Neurobiology of Affiliation*. Cambridge, MA: MIT Press.
Cassirer, E. (1951). *The Philosophy of the Enlightenment*. Princeton, NJ: Princeton University Press.
Chao, L. L., Martin, A., and Haxby, J. V. (1999). Are face-responsive regions selective only for faces? *NeuroReport, 10*: 2945–2950.
Chapados, C. and Levitin, D. J. (2008). Cross-modal interactions in the experience of musical performances: Physiological correlates. *Cognition, 108*(3), 639–651.
Chapin, H., Jantzen, K., Kelso, J.A., Steinberg, F., and Large, E. (2010). Dynamic emotional and neural responses to music depend on performance expression and listener experience. *PLOS ONE, 5*(12): e13812.
Chen, J. L., Penhune, V. B., and Zatorre, R. J. (2008). Listening to musical rhythms recruits motor regions of the brain. *Cerebral Cortex Advance Access*. Oxford: Oxford University Press
Cheney, D. L. and Seyfarth, R. M. (1990). *How Monkeys See the World*. Chicago: University of Chicago Press.
Cheney, D. L. and Seyfarth, R. M. (2007). *Baboon Metaphysics*. Chicago: University of Chicago Press.
Choleris, E., Little, S. R., Mong, J. A., Puram, S. V., Langer, R. and Pfaff, D. W. (2007). Microparticle-based delivery of oxytocin receptor antisense DNA in the medial amygdala blocks social recognition in female mice. *Proceedings of the National Academy of Sciences of the United States, 104*: 4670–4675.
Chomsky, N. (1965). *Aspects of the Theory of Syntax*. Cambridge, MA: MIT Press.
Chomsky, N. (1972). *Language and the Mind*. New York: Harcourt Press.
Chow, I., Brown, S., Poon, M., and Weishaar, K. (2010). A musical template for phrasal rhythm in spoken Cantonese. In *Speech Prosody 2010—Fifth International Conference*.
Clark, A. (1998). *Being There: Putting Brain, Body, and World Together Again*. Cambridge, MA: Bradford Books, MIT Press.
Clarke, E. and Cook, N. (2004). *Empirical Musicology*. Oxford: Oxford University Press.
Clarke, E. (2008). *Ways of Listening: An Ecological Approach to the Perception of Musical Meaning*. Oxford: Oxford University Press.
Cohen, A. and Cohen, N. (1973). Tune evolution as an indicator of traditional musical norms. *Journal of American Folklore, 86*: 37–47.
Cohen, A., Trehub, S., and Thorpe, L. (1989). Effects of uncertainty on melodic information processing. *Perception and Psychophysics, 46*: 18–28.
Cohen, M. A., Evans, K. K., Horowitz, T. S., and Wolfe, J. M. (2011) Auditory and visual memory in musicians and nonmusicians. *Psychonomic Bulletin & Review. 18*: 586–591.
Cohen, M. S., Kooslyn, S. M., Breitter, H. C., DiGirolamo, G. J., Thompson, W. L., Anderson, A. K., et al. (1996). Changes in cortical activity during mental rotation: A mapping study using functional MRI. *Brain, 119*: 89–100.

Conard, N. J. (2009). A female figurine from the basal Aurignacian of Hohle Fels Cave in Southwestern Germany. *Nature, 459*: 248–252.

Conard, N. J., Malina, M., and Munzel, S. C. (2009). New flutes document the earliest musical tradition in southwestern Germany. *Nature, 460*: 737–740.

Conlon, J. M. and Larhammar, D. (2005). The evolution of neuroendocrine peptides. *General and Comparative Endocrinology, 142*(1–2): 53–59.

Connor, R. C. (2007). Dolphin social intelligence: Complex alliance relationships in bottlenose dolphins and a consideration of selective environments for extreme brain size evolution in mammals. *Philosophical Transactions of the Royal Society of London, 362*: 2871–2888.

Cook, N. (1998). *Music: A Very Short Introduction.* Oxford: Oxford University Press.

Cooke, D. (1959, 1964). *The Language of Music.* Oxford: Oxford University Press.

Copland, A. (1937, 1957). *What to Listen for in Music.* New York: McGraw.

Copland, A. (2006) Music and Imagination. Cambridge: Cambridge University Press.

Coppens, Y. (1994). East side story: The origin of human kind. *Scientific American, 270*: 88–95.

Corballis, M. C. and Lea, S.E.G. (1999). *The Descent of Mind.* Oxford: Oxford University Press.

Corballis, M. C. (2002). *From Hand to Mouth.* Princeton: Princeton University Press.

Corballis, M. C. (2004). The origins of modernity: Was autonomous speech the critical factor? *Psychological Review, 111*: 543–52.

Cosmides, L. and Tooby, J. (1992). Cognitive adaptations for social exchange. In J. Barkow, L. Cosmides, and J. Tooby, (eds.), *The Adapted Mind* (pp. 163–228). New York: Oxford University Press.

Cox, A. (2011) Embodying music: principles of the mimetic hypothesis. *Society for Music Theory, 17*: 1–24.

Craig, W. (1918). Appetites and aversions as constituents of instinct. *Biological Bulletin, 34*: 91–107.

Crews, D. (2005). Evolution of neuroendocrine mechanisms that regulate sexual behavior. *TRENDS in Endocrinology and Metabolism, 16*: 354–361.

Crews, D. (2008). Epigenetics and its implications for behavioral endocrinology. *Frontiers in Neuroendocrinology, 29*: 344–357.

Crews, D. (2011). Epigenetic modifications of brain and behavior: Theory and practice. *Hormones and Behavior*, *59*: 393–398.

Crickmore, L. (2003). A re-valuation of the ancient science of harmonics. *Psychology of Music, 31*: 391–403.

Critchley, M. and Critchley, E. A. (1998). *John Hughlings Jackson: Father of English Neurology*. Oxford: Oxford University Press.

Croce, B. (1909, 1983). *Aesthetic.* Boston: Nonpareil Books.

Cross, I. (1999). Is music the most important thing we ever did? In S. W. Yi (ed.), *Music, Mind and Science* (pp. 10–39). Seoul: Seoul National University Press.

Cross, I. (2009a). Music as a communicative medium. *Prehistory of Language, 1*: 113–144.

Cross, I. (2009b). The evolutionary nature of musical meaning. *Musicae Scientiae, Special Issue*: 179–200.

Cross, I. (2010). The evolutionary basis of meaning in music: Some neurological and neuroscientific implications. In F. C. Rose (ed.), *The Neurology of Music* (pp. 1–15). London: Imperial College Press.

Cross, I. and Morley, I. (2008). The evolution of music: theories, definitions and nature of the evidence. In S. Malloch and C. Trevarthen (eds.), *Communicative Musicality* (pp. 61–82). Oxford: Oxford University Press.

Cuddy, L. and Badertscher, B. (1987). Recovery of the tonal hierarchy: Some comparisons across age and levels of musical experience. *Perception and Psychophysics, 41*: 609–620.

Cuddy, L. L. and Lunny C. A. (1995). Expectancies generated by melodic intervals: perceptual judgments of melodic continuity. *Perception and Psychophysics 57*: 451–462.

Cuddy, L. L., Balkwill, L. L., Peretz, I., and Holden, R. R. (2005). Musical difficulties are rare: A study of "tone deafness" among university students. *Annals of the New York Academy of Sciences, 1060*: 311–324.

Cumming, N. (1999). Musical signs and subjectivity: Peircean reflections. *Transactions of the Charles S. Peirce Society, 35*: 437–474.

Cumming, N. (2000). *The Sonic Self: Musical Subjectivity and Signification.* Bloomington: Indiana University Press.

Curley, J. P. and Keverne, E. B. (2005). Genes, brains and mammalian social bonds. *TRENDS in Ecology and Evolution, 20*: 561–567.

Curtis, M. E. and Bharucha, J. J. (2009). Memory and musical expectation for tones in cultural context. *Music Perception, 26*: 365–475.

Cutting, J. E. and Rosner, B. S. (1974). Categories and boundaries in speech and music. *Perception and Psychosis, 16*: 564–570.

Daffner, K. R., Mesulam, M. M., Holcomb, P. J., Calvo, V., Acar, D., Chabrerie, A., et al. (2000). Disruption of attention to novel events after frontal lobe injury in humans. *Journal of Neurology, Neurosurgery and Psychiatry, 68*: 18–24.

Dalla Bella, S., Giguere, J., and Peretz, I. (2007). Singing proficiency in the general population. *Journal of Acoustical Society of America, 121*: 1182–1189.

Dalla Bella, S. and Peretz, I. (2003). Congenital amusia interferes with the ability to synchronize with music. *Annals of the New York Academy of Sciences, 999*: 166–269.

Dallmann, R. and Geissmann, T. (2009). Individual and geographical variability in the songs of wild silvery gibbons on Java, Indonesia. *The Gibbons: Developments in Primatology, Progress and Prospects,* Part 2: 91–110.

Damasio A. R. (1994). *Descartes' Error: Emotion, Reason, and the Human Brain.* New York: Grosset/Putnam.

Damasio, A. R. (1996). The somatic marker hypothesis and the possible functions of the prefrontal cortex. *Philosophical Transactions of the Royal Society of London, 351*: 1413–1420.

Darwin, C. (1859/1958). *The Origin of Species*. New York: Mentor Book.

Darwin, C. (1871/1874). *Descent of Man.* New York: Rand McNally.

Darwin, C. (1872/1998). *The Expression of the Emotions in Man and Animals.* Oxford: Oxford University Press.

Davidson, R. J., Putnam, K. M., and Larson, C. L. (2000). Dysfunction in the neural circuitry of emotion regulations—a possible prelude to violence. *Science, 289*: 591–594.

Davies, J. B. (1978). *The Psychology of Music*. London: Hutchinson.
Davies, S. (1994). *Musical Meaning and Expression*. Ithaca: Cornell University Press.
Dawkins, R. (2004). *A Pilgrimage to the Dawn of Life*. London: Phoenix.
Daynes, H. (2010). Listeners perceptual and emotional responses to tonal and atonal music. *Psychology of Music, 39*: 468–502.
Decety, J., Perani, D., and Jeannerod, M. (1994). Mapping motor representations with positron emission tomography. *Nature, 371*: 600–602.
Dehaene, S. (1997). *The Number Sense*. Oxford: Oxford University Press.
Dehaene, S., Izard, V., Pica, P., and Spelke, E. (2006). Core knowledge of geometry in an Amazonian Indigene group. *Science, 311*: 381–384.
Delson, E. and Harvati, K. (2006). Return of the last Neanderthal. *Nature, 443*: 262–263.
Delza, S. (1996). *The T'ai-Chi Ch'uan experience*. Albany: State University of New York Press.
Demorest, S. M., Morrison, S. J., Stambaugh, L. A., Beken, M., Richards, T. L., and Johnson, C. (2010). An fMRI investigation of the cultural specificity of music memory. *SCAN, 5*: 282–291.
Dennett, D. (1987). *The Intentional Stance*. Cambridge, MA: MIT Press.
Denton, D. (1982). *The Hunger for Salt*. Berlin: Springer-Verlag.
Desimone, R. (1996). Neural mechanisms for visual memory and their role in attention. *Proceedings of the National Academy of Sciences, 93*: 13494–13499.
Desimone, R., Albright, T. D., Gross, C. G., and Bruce, C. (1984). Stimulus-selective properties of inferior temporal neurons in the macaque. *Journal of Neuroscience, 8*: 2051–2062.
Desmurget, M., Reilly, K. T., Richard, N., Szathmari, A., Mottolese, C., and Sirigu, A. (2009). Movement intention after parietal cortex stimulation in humans. *Science, 324*: 811–813.
Deutsch. D. (1999). *The Psychology of Music*, 2nd ed. San Diego: Academic Press.
Deutsch, D. (2007). Music Perception. *Frontiers of Bioscience*, 1–10.
Devi, R. (1962). *Dances of India*. Calcutta: Susil Gupta.
De Vries, G. J., Duetz, W., Ruud, B. M., Van Heerikhuize, J., and Vreeburg, J. T. M. (1986). Effects of androgens and estrogens on the vasopressin and oxytocin innervations of the adult rat brain. *Brain Research, 399*: 296–302.
De Vries, A. C., DeVries, M. B., Taymans, S., and Carer, C. S. (1995). Modulation of pair bonding in female voles by corticosterone. *Proceedings of the National Academy of the Sciences, 92*: 7744–7748.
De Vries, G. J. (2008). Sex differences in vasopressin and oxytocin innervations of the brain. *Progress in Brain Research, 170*: 17–25.
De Vries, G. J. and Simerly, R. B. (2002). Anatomy, development and function of sexually dimorphic neural circuits in the mammalian brain. In D. W. Pfaff, et al. (eds.), *Hormones, Brain and Behavior* (pp. 137–191). New York: Elsevier.
Dewey, J. (1895). The theory of emotion. *Psychological Review, 2*(1): 13–32.
Dewey, J. (1896). The reflex arc concept in psychology. *Psychological Review, 3*: 357–370.
Dewey, J. (1929/1960). *The Quest for Certainty*. New York: Capricorn Books.
Dewey, J. (1925/1989). *Experience and Nature*. La Salle: Open Court.
Dewey, J. (1934/1958). *Art as Experience*. New York: Capricorn Books.

Dewey, J. (1938). *Logic: The Theory of Inquiry*. New York: Holt, Rinehart.

Diamond, A. (2001). A model system for studying the role of dopamine in the prefrontal cortex during early development of humans: Early and continuously treated phenylketonuria. In C. A. Nelson and M. Luciana (eds.), *Handbook of Developmental Cognitive Neuroscience* (pp. 433–472). Cambridge, MA: MIT Press.

Dickinson, A. (1980). *Contemporary Animal Learning*. Cambridge: Cambridge University Press.

Diderot, D. (1755, 1964). The Encyclopedia. In *Rameau's Nephew and Other Works*. New York: Library of Liberal Arts.

Dipert, R. R. (1983). Meyer's emotion and meaning in music: A sympathetic critique of its central claims. *Journal of the Michigan Music Theory Society, 6*: 3–18.

Dobzhansky, T. C. (1962). *Mankind Evolving*. New Haven: Yale University Press.

Domes, G., Heinrichs, M., Glascher, J., Buchel, C., Braus, D. F., and Herpertz, S. C. (2007). Oxytocin attenuates amygdala responses to emotional faces regardless of valence. *Biological Psychiatry, 62*: 1187–1190.

Don, A., Schellenberg, E. G., and Rourke, B. P. (1999). Music and language skills of children with Williams syndrome. *Child Neuropsychology, 5*: 154–170.

Donald, M. (1991). *Origins of the Modern Mind*. Cambridge, MA: Harvard University Press.

Donald, M. (1998). Hominid enculturation and cognitive evolution. In C. Renfrew and C. Scarre (eds.), *Cognition and Material Culture: The Archaeology of Symbolic Storage* (pp. 7–17). Cambridge, UK: McDonald Institute for Archaeological Research.

Donald, M. (2001). *A Mind So Rare: The Evolution of Human Consciousness*. New York: Norton.

Donaldson, Z. R. and Young, L .J. (2008). Oxytocin, vasopressin, and the neurogenetics of sociality. *Science, 322*: 900–904.

Donaldson, Z. R., Kondrashov, F. A., Putnam, A., Bai, Y., Stoinski, T. L., Hammock, E. A., et al. (2008). Evolution of a behavior-linked microsatellite-containing element in the 5' flanking region of the primate AVPR1A gene. *BioMed Central Evolutionary Biology, 8*: 18.

Dowling, W. J. and Harwood, D. L. (1986). *Music Cognition* (Vol. 19986). New York: Academic Press.

Dowling, W. J., Tillmann, B., and Ayers, D. F. (2002). Memory and the experience of hearing music. *Music Perception, 19*: 249–276.

Downing, T. (1995). *Music and the Origins of Language*. Cambridge: Cambridge University Press.

Dunbar, R.I.M. (1996). *Grooming, Gossip and the Evolution of Language*. Cambridge, MA: Harvard University Press.

Dunbar, R.I.M. (1998). The social brain hypothesis. *Evolutionary Anthropology, 6*: 178–190.

Dunbar, R.I.M. (2003). The social brain. *Annual Review of Anthropology, 32*: 163–181.

Dunbar, R.I.M. and Shultz, S. (2007). Evolution in the social brain. *Science, 317*: 1344–1347.

Dunbar, R. (2010). *How Many Friends Does One Person Need?* Cambridge, MA: Harvard University Press.

Dunlop, R. A., Noad, M. J., Cato, D. H., and Stokes, D. (2007). The social vocalization repertoire of east Australian migrating humpback whales. *Journal of the Acoustical Society of America, 122*: 2893–2905.

Dupre, J. (1981). Natural kinds and biological taxa. *Philosophical Review, 90*: 66–90.

Dykins, E. M., Rosner, B. A., Ly, T., and Sagun, J. (2005). Music and anxiety in Williams syndrome: A harmonious or discordant relationship. *American Journal of Mental Retardation, 110*: 346–258.

Eco, U. (1976). *A Theory of Semiotics*. Bloomington: Indiana University Press.

Edelman, S. (1997). Curiosity and exploration. Retrieved from http://www.csun.edu/~vcpsy00h/students/explore.htm on May 11, 2005.

Eerola T., Louhivuori J., and Lebaka, E. (2009). Expectancy in Sami Yoiks revisited: The role of data driven and schema-driven knowledge in the formation of melodic expectations. *Musicae Scientiae, 13*: 231–272.

Eichenbaum, J. and Cohen, N. J. (2001). *From Conditioning to Conscious Recollection*. Oxford: Oxford University Press.

Ekman, P. (1972). Universals and cultural differences in facial expressions of emotion. In J. Cole (ed.), *Nebraska Symposium on Motivation, 1971*. Lincoln: University of Nebraska Press.

Eldridge, N. (1985). *Unfinished Synthesis*. Oxford: Oxford University Press.

Eldridge, N. (1999). *The Pattern of Evolution*. New York: W. H. Freeman and Co.

Elliott, R., Newman, J. L., Longe, O. A., and Deakin, J.F.W. (2003). Differential response patterns in the striatum and orbitofrontal cortex to financial reward in humans: A parametric functional magnetic resonance imaging study. *Journal of Neuroscience, 23*: 303–307.

Eliot, T. S. (1943). *Four Quartets*. New York: Harcourt Press.

Elster, J. (2000). *Ulysses Unbound*. Cambridge: Cambridge University Press.

Emery, N. J. (2000). The eyes have it: The neuroethology, function and evolution of social gaze. *Neuroscience and Biobehavioral Reviews, 24*: 581–604.

Emery, N. J. and Amaral, D. G. (2000).The role of the amygdala in primate social cognition. In R. D. Lane and L. Nadel (eds.), *Cognitive Neuroscience of Emotion* (pp. 156–191). New York: Oxford University Press.

Enard, W., Gehre, S., Hammerschmidt, K., Holter, S. M., Blass, T., Somel, M., et al. (2009). A humanized version of Foxp2 affects cortico-basal ganglia circuits in mice. *Cell, 137*: 961–971.

Engel, A. and Keller, P. E. (2011). The perception of musical spontaneity in impoverished and imitated jazz performances. *Frontiers in Psychology, 2*: 1–12.

Epstein, J. (2008). *Fred Astaire*. New Haven: Yale University Press.

Evers, S. and Suhr, B. (2000). Changes of the neurotransmitter serotonin but not of hormones during short time music perception. *European Archives of Psychiatry and Clinical Neuroscience, 250*: 144–147.

Faber, M. (1983). *Mozart*. New York: Vintage Press.

Fadiga, L., Craighero, L., and Ausilio, A. D. (2009). Broca's area in language, action, and music. *The Neurosciences and Music III– Disorders and Plasticity, 1168*: 448–458.

Falk, D. (1983). Cerebral cortices of east African early hominids. *Science, 221*: 1072–1074.

Falk, D. (2000). Hominid brain evolution and the origins of music. In N. L. Wallin, B. Merker, and S. Brown (eds.), *The Origins of Music* (pp. 197–216). Cambridge, MA: MIT Press.

Farah, M. J. (1984). The neurobiological basis of visual imagery: A componential analysis. *Cognition, 18*: 245–272.

Faure, A., Richard, J. M., and Berridge, K. C. (2010). Desire and dread from the nucleus accumbens: Corticol glutamate and subcortical GABA differentially generate motivation and hedonic impact in the rat. *PlosOne, 5*: e11223.

Feder, S. (2004). *Gustav Mahler*. New Haven: Yale University Press.

Ferris, C. F. (2008). Functional magnetic resonance imaging and the neurobiology of vasopressin and oxytocin. *Progress in Brain Research, 170*: 305–320.

Finlayson, C. (2004). *Neanderthals and Modern Humans*. Cambridge: Cambridge University Press.

Finlayson, C., Pacheco, F. G., Rodríguez-Vidal, J., Fa, D. A., López, J. M. G., Perez, A. S., et al. (2006). Late survival of Neanderthals at the southernmost extreme of Europe. *Nature, 443*: 850–853.

Fiorillo, C. D., Tobler, P. N., and Schultz, W. (2003). Discrete coding of reward probability and uncertainty by dopamine neurons. *Science, 299*: 1898–1902.

Fischer, J. M. (2011). *G. Mahler*. New Haven: Yale University Press.

Fisch, M. H. (1986). *Peirce, Semiotic and Pragmatism: Essays by Max H. Fisch*. K. L. Ketner and J. W. Kloesel (eds.), Bloomington: Indiana University Press.

Fisher, R. A. (1930). *The Genetic Theory of Natural Selection*. Oxford: Clarendon Press.

Fiske, A. P. (1991). *Structures of Social Life*. New York: Free Press.

Fitch, W. T. (2000). The evolution of speech: A comparative review. *Trends in Cognitive Sciences, 4*: 258–267.

Fitch, W. T. (2006). The biology and evolution of music: A comparative perspective. *Cognition, 100*: 173–215.

Fitch, W. T. (2009). Birdsong normalized by culture. *Nature, 459*: 519–520.

Fitzsimons, J. T. (1979). *The Physiology of Thirst and Sodium Appetite*. Cambridge: Cambridge University Press.

Fitzsimons, J. T. (1999). Angiotensin, thirst and sodium appetite. *Physiological Reviews, 76*: 583–687.

Foley, R. (1996). An evolutionary and chronological framework for human social behaviour. *Proceedings of the British Academy, 88*: 95–117.

Foley, R. (2001). In the shadow of the modern synthesis? *Evolutionary Anthropology, 10*: 5–14.

Foley, R. (2006). The emergence of culture in the context of hominin evolutionary patterns. In S. C. Levinson and P. Jaisson (eds.), *Evolution and Culture* (pp. 53–78). Cambridge, MA: MIT Press.

Foley, R. and Lahr, M. M. (2003). On stony ground: Lithic technology, human evolution and the emergence of culture. *Evolutionary Anthropology, 12*: 109–122.

Fortune, E. S., Rodriguez, C., Li, D., Ball, G. F., and Coleman, M. J. (2011). Neural mechanisms for the coordination of duet singing in Wrens. *Science, 334*: 666–670.

Foster, R. G. and Kreitzman, L. (2004). *Rhythms of Life*. New Haven: Yale University Press.

Fox, N. A., Nichols, K. E., Henderson, H. A., Rubin, K., Schmidt, L., Hamer, D., et al. (2005). Evidence for a gene-environment interaction in predicting behavioral inhibition in middle childhood. *Psychological Science, 16*: 1770–17784.

Franklin, M. R., Moore, C. S., Yip, C., Jonides, J., Rattray, K., and Moher, J. (2008). The effects musical training on verbal memory. *Psychology of Music, 36*: 353–365.

Frederickson, B. L. (2004). The broaden-and-build theory of positive emotions. *Philosophical Transactions of the Royal Society of London, 359*: 1367–1377.

Freedland, M. (1972). *Al Jolson*. London: W. H. Allen and Co.

Frey, S. H. (2007). What puts the how in where? Tool use and the divided visual streams hypothesis. *Cortex, 43*: 368–375.

Fried, I., Wilson, C. L., Morrow, J. W., Cameron, K. A., Behnke, E. D., Ackerson, L. C., et al. (2001). Increased dopamine release in the human amygdala during performance of cognitive tasks. *Nature Neuroscience, 4*: 201–206.

Friederici, A. D., Wang, Y., Herrmann, C. S., Maess, B., and Oertel, U. (2000). Localization of early syntactic processes in frontal and temporal cortical areas: A magnetoencephalographic study. *Human Brain Mapping, 11*: 1–11.

Friederici, A. D. (2002). Towards a neural basis of auditory sentence processing. *TRENDS in Cognitive Sciences, 6*: 78–84.

Fristrup, K. M., Hatch, L. T., and Clark, C. W. (2003). Variation in humpback whale song length in relation to low-frequency sound broadcasts. *Journal of the Acoustical Society of America, 113*: 3411–3424.

Frith, C. D. (2007). The social brain? *Philosophical Transactions of the Royal Society of London, 362*: 671–678.

Frith, C. D. and Wolpert, D. (2003). *The Neuroscience of Social Interaction*. Oxford: Oxford University Press.

Fromm, E. (1973). *The Anatomy of Human Destructiveness*. New York: Holt, Rinehart and Winston.

Gage, F. H. (1998). Stem-cells of the central nervous system. *Current Opinions in Neurobiology, 8*: 671–676.

Galaburda, A. M., Schmitt, J. E., Atlas, S. W., Eliez, S., Bellugi, U., and Reiss, A. L. (2001). Dorsal forebrain anomaly in Williams syndrome. *Archives of Neurology, 58*: 1865–1869.

Galileo (1610/1957). The starry messenger. In S. Drake (ed.), *Discoveries and Opinions of Galileo*. New York: Double Day.

Galison, P. (1988). History, philosophy and the central metaphor. *Science in Context, 3*: 197–212.

Gallagher, M. and Holland, F. C. (1994). The amygdala complex: Multiple roles in associative learning and emotion. *Proceedings of the National Academy of Sciences, 91*: 11771–11776.

Gallagher, S. (2005). *How the Body Shapes the Mind*. Oxford: Oxford University Press.

Gallese, V., Fadiga, L., Fogassi, L., and Rizzolatti, G. (1996). Action recognition in the premotor cortex. *Brain, 119*: 593–609.

Gallese, V. and Goldman, A. (1998). Mirror neurons and the simulation theory of mind-reading. *Trends in Cognitive Science, 2*: 493–501.

Gallistel, C. R. (1980). *The Organization of Action: A New Synthesis*. Hillsdale: Lawrence Erlbaum.
Gallistel, C. R. (1990). *The Organization of Learning*. Cambridge, MA: MIT Press.
Gallistel, C. R., Gelman, R. M., and Cordes, S. (2005). The cultural and evolutionary history of the real numbers. In S. Levinson, and P. Jaisson (eds.), *Culture and Evolution* (Vol. 17, pp. 247–274). Cambridge, MA: MIT Press.
Gamble, C., Davies, W., Pettitt, P., and Richards, M. (2004). Climate change and evolving human diversity in Europe during the last glacial. *Philosophical Transactions of the Royal Society of London B: Biological Science, 359*: 243–253.
Garamszegi, L. Z., Pavlova, D. Z., Eens, M., and Moller, A. P. (2007). The evolution of song in female birds in Europe. *Behavioral Ecology, 18*: 86–96.
Garcia, J., Hankins, W. G., and Rusiniak, K. W. (1974). Behavioral regulation of the milieu interne in man and rat. *Science, 185*: 824–831.
Gaser, C. and Schlaug, G. (2003). Brain structures differ between musicians and non-musicians. *Journal of Neuroscience, 23*(27): 9240–9245.
Gazzaniga, M. S. (1995, 2000). *The New Cognitive Neurosciences*. Cambridge, MA: MIT Press.
Gazzaniga, M. S. (1998). *The Mind's Past*. Berkeley: University of California Press.
Geissmann, T. (2000). Gibbon songs and human music from an evolutionary perspective. In N. L. Wallin, B. Merker, and S. Brown (eds.), *The Origins of Music* (pp. 103–123). Cambridge, MA: MIT Press.
Geissmann, T. (2002). Duet-splitting and the evolution of gibbon songs. *Biological Review, 77*: 57–76.
Geissmann, T., Bohlen-Eyring, S., and Heuck, A. (2005). The male song of the Javan silvery gibbon. *Contributions to Zoology, 74*: 1–25.
Gelman, R. and Gallistel, C. R. (1978). *The Child's Understanding of Number*. Cambridge, MA: Harvard University Press.
Gelman, S. A. (2003). *The Essential Child*. Oxford: Oxford University Press.
Geschwind, N. (1974). *Selected Papers on Language and the Brain*. Boston: Reidel Publishing.
Gibbs, R. W. (2006). *Embodiment and Cognitive Science*. Cambridge: Cambridge University Press.
Gibson, J. J. (1966). *The Senses Considered as Perceptual Systems*. New York: Houghton-Mifflin.
Gibson, J. J. (1979). *The Ecological Approach to Visual Perception*. Boston: Houghton Mifflin.
Gibson, K. R. and Ingold, T. (eds.) (1993). *Tools, Language and Cognition in Human Evolution*. Cambridge: Cambridge University Press.
Gigerenzer, G. (1991). From tools to theories: A heuristic of discovery in cognitive psychology. *Psychological Review, 98*: 254–267.
Gigerenzer, G. (2000). *Adaptive Thinking: Rationality in the Real World*. New York: Oxford University Press.
Gigerenzer, G. (2007). *Gut Feelings*. New York: Viking Press.
Gimpl, G. and Fahrenholz, F. (2001). The oxytocin receptor system: Structure, function and regulation. *Physiology Review, 81*: 629–683.

Gjerdingen, R. O. (1989). Meter as a mode of attending: A network simulation of attentional rhythmicity in music. *Integral, 3*: 67–91.
Gjerdingen, R. O. (1990). Categorization of musical patterns by self-organizing neuronlike networks. *Music Perception, 8*: 339–370.
Gjerdingen, R. O. (1999). An experimental music theory? In N. Cook and M. Everist (eds.), *Rethinking Music* (pp. 161–170). Oxford: Oxford University Press.
Gjerdingen, R. (2007). The psychology of music. In T. Christensen (ed.), *The Cambridge History of Western Music Theory* (pp. 956–981). Cambridge: Cambridge University Press.
Gjerdingen, R. O. (2010). Leonard Meyer remembered. *Aesthetics, online*: http://www.aesthetics-online.org/memorials/index.php?memorials_id=33.
Gjerdingen, R. O. (2009). Meyer and 'Music Usage'. Unpublished paper.
Goethe, J.F.V. (1790, 2009). *The Metamorphosis of Plants*. Cambridge, MA: MIT Press.
Goldin-Meadow, S. (1999). The role of gesture in communication and thinking. *Trends in Cognitive Sciences, 3*: 419–429.
Goldin-Meadow, S., Cook, S. W., and Mitchell, Z. A. (2009). Gesturing gives children new ideas about math. *Psychological Science, 20*: 267–272.
Goldstein, D. G. and Gigerenzer, G. (2002). Models of ecological rationality: The recognition heuristic. *Psychological Review, 109*: 75–90.
Goldstein, M. H., King, A. P., and West, M. J. (2003). Social interaction shapes babbling: Testing parallels between birdsong and speech. *Proceedings of the National Academy of Sciences, 100*: 8030–8035.
Goodale, M. A. and Milner, A. D. (1992). Separate visual pathways for perception and action. *Trends in Neuroscience, 15*: 20–25.
Goodman, N. (1955, 1978). *Fact, Fiction and Forecast*. New York: Bobbs-Merrill.
Goodman, N. (1968). *Languages of Art*. New York: Bobbs-Merrill.
Goodson, J. L. (2008). Nonapeptides and the evolutionary patterning of sociality. *Progress in Brain Research, 170*: 3–15.
Goodson, J. L. and Bass, A. H. (2000). Forebrain peptides modulate sexually polymorphic vocal circuitry. *Nature, 403*: 769–772.
Gosselin, N., Samson, S., Adolphs, R., Noulhiane, M., Roy, M., Hasboun, D., et al. (2006). Emotional responses to unpleasant music correlates with damage to the parahippocampal cortex. *Brain, 129*: 2585–2592.
Gould, E., Beylin, A., Tanapat, P., Reeves, A., and Shors, T. J. (1999). Learning enhances adult neurogenesis in the hippocampal formation. *Nature Neuroscience, 2*: 260–265.
Gould, S. J. (1989). Church, Humboldt, and Darwin: The Tension and Harmony of Art and Science. In F. Kelly (ed.), *Frederic Edwin Church* (pp. 94–107). Washington: Smithsonian Institution Press.
Gould, S. J. (2002). *The Structure of Evolutionary Theory*. Cambridge, MA: Harvard University Press.
Gould, S. J. and Eldridge, N. (1977). Punctuated equibria: The tempo and mode of evolution reconsidered. *Paleobiology, 3*: 115–151.
Goy, R. W. and McEwen, B. S. (1980). *Sexual Differentiation of the Brain*. Cambridge, MA: MIT Press.
Grahn, J. A. (2009). The role of the basal ganglia in beat perception. *The Neurosciences and Music III–Disorders and Plasticity, 1169*: 35–25.

Grahn, J. A. and Brett, M. (2007). Rhythm and beat perception in motor areas of the brain. *Journal of Cognitive Neuroscience, 19*: 893–906.

Grahn, J. A., and Brett, M. (2009). Impairment of beat-based rhythm discrimination in Parkinson's disease. *Cortex, 45*: 54–61.

Grahn, J. A. and Rowe, J. B. (2009). Feeling the beat: premotor and striatal interactions in musicians and nonmusicians during beat perception. *Journal of Neuroscience, 29*: 7540–7548.

Grauer, V. C. (2006). Echoes of our forgotten ancestors. *The World of Music, 48*: 5–39.

Gray, P. M., Krause, B., Atema, J., Payne, R., Krumhansl, C. and Baptista, L. (2001). The music of nature and the nature of music. *Science, 291*: 52–54.

Graybiel, A. M. (1998). The basal ganglia and chunking of action repertoires. *Neurobiology of Learning and Memory, 70*: 119–136.

Graybiel, A. M., Aosaki, T., Flaherty, A. W., and Kimura, M. (1994). The basal ganglia and adaptive motor control. *Science, 265*: 1826–1831.

Green, R. E., Krause, J., Briggs, A. W., Maricic, T., Senzel, U., Kircher, M., et al. (2010). A draft sequence of the Neanderthal genome. *Science, 328*: 710–722.

Green, R. E., Krause, J., Ptak, S. E., Briggs, A. W., Ronan, M. T., Simons, J. F., et al. (2006). Analysis of one million base pairs of Neanderthal DNA. *Nature, 444*: 330–336.

Grice, P. (1957). Meaning. *Philosophical Review, 66*: 377–388.

Griffin, D. R. (1958). *Listening in the Dark*. New Haven: Yale University Press.

Gritten, A. and King, E. (2006). *Music and Gesture*. Surrey: Ashgate Publishing.

Grocke, D. and Wigram, T. (2007). *Receptive Methods in Music Therapy*. London: Jessica Kingley.

Gross, C. G. (1999). *Brain, Vision and Memory*. Cambridge, MA: MIT Press.

Grossman, T., Oberecker, R., Koch, S. P., and Friederici, A. D. (2010). The developmental origins of voice processing in the human brain. *Neuron, 65*: 852–858.

Grossmann, K. E. and Grossmann, K. (2003). Universality of human social attachment as an adaptive process. In S. C. Carter, et al. (eds.), *Attachment and Bonding*. Cambridge, MA: MIT Press.

Gunz, P., Neubauer, S., Maureille, B., and Hublin, J. (2010). Brain development after birth differs between Neanderthals and modern humans. *Current Biology, 20*: 1.

Gurney, M. E. (1982). Behavioral correlates of sexual differentiation in the Zebra Finch song system. *Brain Research, 231*: 153–172.

Gurney, M. E. and Konishi, M. (1980). Hormone-induced sexual differentiation of brain and behavior in Zebra Finches. *Science, 208*: 1380–1383.

Guthrie, W.K.C. (1955). *The Greeks and Their Gods*. Boston: Beacon Press.

Haas, B. W., Mills, D. Yam, A., Hoeft, F., Bellugi, U., and Reiss, A. (2009). Genetic influences on sociability: heightened amygdala reactivity and event-related responses to positive social stimuli in Williams syndrome. *Journal of Neuroscience, 29*: 1132–1139.

Hacking, I. (1964). *Logic of Statistical Inference*. Cambridge: Cambridge University Press.

Hacking, I. (1975). *The Emergence of Probability*. Cambridge: Cambridge University Press.

Hacking, I. (1999). *The Taming of Chance.* Cambridge: Cambridge University Press.

Haidt, J. (2001). The emotional dog and its rational tail. *Psychological Review, 108*: 814–824.

Haidt, J. and Morris, J. P. (2009). Finding the self in self-transcendent emotions. *Proceedings of the National Academy of Sciences of the United States, 106*: 7687–7688.

Hamilton, R. H., Pascual-Leone, A., and Schlaug, G. (2004). Absolute pitch in blind musicians. *NeuroReport, 15*: 803–806.

Hammock, E. A. and Young, L. J. (2004). Functional microsatellite polymorphism associated with divergent social structure in vole species. *Molecular Biology and Evolution, 21*: 1057–1063.

Han, S., Sundararajan, J., Bowling, D. L., Lake, J., and Purves, D. (2011). Co-variation of tonality in the music and speech of different cultures. *PLoS ONE, 6*(5): e20160. Doi:10.1371/journal.pone.00201600.

Handel, S., Todd, S. K., and Zoidis, A. M. (2009). Rhythmic structure in humpback whale songs: Preliminary implications for song production and perception. *Journal of the Acoustical Society of America, 6*: EL225–EL230.

Handy, T. C., Grafton, S. T., Schroff, N. M., Ketay, S., and Gazzaniga, M. S. (2003). Graspable objects grab attention when the potential for action is recognized. *Nature Neuroscience, 6*: 421–427.

Hannon, E. E. and Trehub, S. E. (2005). Tuning in to musical rhythms: infants learn more readily than adults. *Proceedings of the National Academy of Sciences of the United States, 102*: 12639–12643.

Hannon E. E. and Trainor, L. J. (2007). Music acquisition. *TRENDS in Cognitive Science, 11*: 485–491.

Hanslick, E. (1854). *The Beautiful in Music* (translated by Gustav Cohen), 7th ed. (1855) London: Novello, 1891.

Hanson, N. R. (1971). *Observation and Explanation*. New York: Harper Press.

Hanson, N. R. (1958, 1972). *Patterns of Discovery.* Cambridge: Cambridge University Press.

Hari, R., Forss, N., Avikainen, S., Kirverskari, E., Salenius, S., and Rizzolatti, G. (1998). Activation of human primary cortex during action observation: A neuromagnetic study. *Proceedings of the National Academy of Sciences, 95*: 15061–15065.

Hartshorne, C. (1973). *Born to Sing*. Bloomington: Indiana University Press.

Hatten, R. S. (2004). *Interpreting Musical Gesture*: Bloomington: Indiana University Press.

Haueisen, J. and Knosche, T. R. (2001). Involuntary motor activity in pianists evoked by music perception. *Journal of Cognitive Neuroscience, 13*: 786–792.

Hauk, O., Johnsrude, I., and Pulvermuller, F. (2004). Somatotopic representation of action words in human motor and premotor cortex. *Neuron, 41*: 301–307.

Hauser, M. D. and Konishi, M. (1999). *The Design of Animal Communication.* Cambridge, MA: MIT Press.

Heaton, P. (2009). Assessing musical skills in autistic children who are not savants. *Philosophical Transactions of the Royal Society of London B, 364*: 1443–1447.

Hebb, D. O. (1949). *The Organization of Behavior*. New York: Wiley.

Hebert, D. G. (2007). Bruno Nettle's the study of ethnomusicology: thirty-one issues and concepts: An essay review. *International Journal of Education and the Arts, 8*, Review 2.

Heelan, P. A. (1983). *Space Perception and the Philosophy of Science*. Berkeley: University of California Press.

Heelan P. A. and Schulkin, J. (1998). Hermeneutical philosophy and pragmatism: A philosophy of the science. *Synthese, 115*: 269–302.

Heidegger, M. (1927, 1962). *Being and Time* (Translated by J. Macquarrie and E. Robinson). New York: Harper & Row.

Helmholtz, H. (1873). *Popular Lectures in Scientific Subjects* (Translated by E. Atkinson with introduction by J. Tyndale). London: Longmans Green.

Herbert, J. and Schulkin, J. (2002). Neurochemical coding of adaptive responses in the limbic system. In D. Pfaff (ed.), *Hormones, Brain and Behavior* (pp. 659–689). New York: Elsevier Press.

Herman, E., Call, J., Hernadez-Lioreda, M. V., Hare, B., and Tomasello, M. (2007). Humans have evolved specialized skills of social cognition. *Science, 317*(5843): 1360–1366.

Hillecke, T., Nickel, A., and Bolay, H. V. (2005). Scientific perspectives on music therapy. German Center for Music Therapy Research, and Outpatient Department, University of Applied Sciences Heidelberg, Germany. *Annals of the New York Academy of Sciences, 1060*: 271–282.

Hofer, M. A. (1973). The role of nutrition in the physiological and behavioral effects of early maternal separation on infant rats. *Psychosomatic Medicine, 35*: 350–359.

Hollander, E. Novotny, S., Hanratty, M., Yaffe, R., DeCaria, C. M., Aronowitz, B. R., et al. (2003). Oxytocin infusion reduces repetitive behaviors in adults with autistic and Asperger's disorders. *Neuropsychopharmacology, 28*: 193–198.

Holliday, R. (2006). Epigenetics: A historical overview. *Epigenetics, 1*: 76–80.

Holliday, R. and Ho, T. (1998). Evidence for gene silencing by endogenous DNA methylation. *Proceedings of the National Academy of Sciences, 95*: 8727–8732.

Horgen, P. (2010). Receptive music therapy. In M. E. Horowitz and S. Sondheim (eds.), *Sondheim on Music*. New York: Scarecrow Press.

Houser, N. and Kloesel, C. (1992, 1998). *The Essential Peirce, Volume 1 (1867–1893)*. Bloomington: Indiana University Press.

Houser, N., Eller, J. R., Lewis, A. C., De Tienne, A., Clark, C. L., and Bront Davis, D. (1998). *The Essential Peirce, Volume 2 (1893–1913)*. Bloomington: Indiana University Press.

Howells, W. W. (1976). Explaining modern man. *Journal of Human Evolution, 5*: 477–495.

Hubel, D. H. and Wiesel, T. N. (2005). *Brain and Visual Perception: The Story of a 25-Year Collaboration*. Oxford: Oxford University Press.

Humbolt, W. von (1836, 1971). *Linguistic and Intellectual Development*. Philadelphia: University of Pennsylvania Press.

Humphrey, D. (1959). *The Art of Making Dancers*. New York: Grove Press.

Humphrey, N. (1976). The social function of intellect. In P.P.G. Bateson and R. A. Hinde (eds.), *Growing Points in Ethology* (pp. 307–317). Cambridge: Cambridge University Press.

Humphrey, N. (2007). The society of selves. *Philosophical Transactions of the Royal Society of London B, 362*: 745–754.

Huron, D. Leonard Meyer, parts I, II, III. Online from: http://www.music-cog.ohio-state.edu/Music829D/Notes/Meyer1.html.

Huron, D. (2001). Is music an evolutionary adaptation? *Annals of the New York Academy of Sciences, 930*: 43–61.

Huron, D. (2005). The plural pleasures of music. In J. Sundberg and W. Brunson (eds.), *Proceedings of the 2004 Music and Music Science Conference* (pp. 1–13). Stockholm: Kungliga Musikhögskolan and KTH (Royal Institute of Technology).

Huron, D. (2008). *Sweet Anticipation: Music and the Psychology of Expectation*. Cambridge, MA: MIT Press.

Huron, D. (2008). Lost in music. *Nature, 453*: 456–457.

Hutchinson, A. (1977). *Labanotation*. London: Taylor and Francis.

Hyde, K. L., Lerch, J. P., Zatorre, R. J., Griffiths, T. D., Evans, A. C., and Peretz, I. (2007). Cortical thickness in congenital amusia: When less is better than more. *Journal of Neuroscience, 27*: 13028–13032.

Insel, T. R. (2010). The challenge of translation in social neuroscience: A review of oxytocin, vasopressin, and affiliative behavior. *Neuron, 65*: 768–779.

Insel, T. R. and Young, L. J. (2001). The neurobiology of attachment. *Neuroscience, 2*: 129–136.

Iversen, J. R., Repp, B. H., and Patel, A. D. (2009). Top-down control of rhythm perception modulates early auditory responses. *Annals of the New York Academy of Sciences, 1169*: 58–73.

Jablonka, E. and Lamb, M. J. (1995). *Epigenetic Inheritance and Evolution*. Oxford: Oxford University Press.

Jackendoff, R. and Lerdahl, F. (2006). The capacity for music: What is it, and what's special about it? *Cognition, 100*: 33–72.

Jackson, J. H. (1884, 1958). Evolution and dissolution of the nervous system. In J. Taylor (ed.), *Collected Works of John Hughlings Jackson (1958)*. Volume 11. London: Staples Press.

Jackson, P. L. and Decety, J. (2004). Motor cognition: A new paradigm to study self-other interactions. *Current Opinion in Neurobiology, 14*: 259–263.

James, J. (1993). *The Music of the Spheres*. New York: Grove Press.

James, W. (1887). Some human instincts. *Popular Science Monthly, 31*: 160–176.

James, W. (1890, 1952). *Principles of Psychology* (Vol. 1 and 2). New York: Henry Holt.

Janata, P. (2005). Brain networks that track musical structure. *Annals of the New York Academy of Sciences, 1060*: 111–125.

Janata, P. and Grafton, S. T. (2003). Swinging in the brain: shared neural substrates for behaviors related to sequencing and music. *Nature Neuroscience, 6*: 682–687.

Jarvis, E. D. and Nottebohm, F. (1997). Motor-driven gene expression. *Proceedings of the National Academy of Sciences of the United States: Neurobiology, 94*: 4097–4102.

Jaspers, K. (1913, 1997). *General Psychopathology*. Baltimore: Johns Hopkins University Press.

Jeannerod, M. (1997). *The Cognitive Neuroscience of Action*. Oxford: Blackwell.

Jeannerod, M. (1999). To act or not to act: perspectives on the representation of action. *Quarterly Journal of Experimental Psychology, 52*: 1–29.

Jentschke, S., Koelsch, S., Sallat, S., and Friederici, A.D. (2008). Children with specific language impairment also show impairment of music-syntactic processing. *Journal of Cognitive Neuroscience, 20*: 1940–1951.

Johanson, D. C. and Edey, M. (1981). *Lucy: The Beginnings of Humankind*. New York: Simon & Schuster.

Johnson, M. L. (1987, 1990). *The Body in the Mind*. Chicago: University of Chicago Press.

Johnson, M. L. (1993). *Moral Imagination*. Chicago: University of Chicago Press.

Johnson, M. L. (2007). *The Meaning of the Body*. Chicago: University of Chicago Press.

Johnson, M. L. and Larson, S. (2003). "Something in the Way She Moves"—Metaphors of Musical Motion. *Metaphor and Symbol 18*: 63–84.

Johnson-Frey, S. H., Maloof, F. R., Newman, Norlund, R., Farrer, C., Inati, S., et al. (2003). Actions or hand-object interactions? Human inferior frontal cortex and action observation. *Neuron, 39*: 1053–1058.

Johnson-Laird, P. N. (2002). Peirce, logic diagrams, and the elementary operations of reasoning. *Thinking and Reasoning, 8*: 69–95.

Johnston, J. B. (1923). Further contributions to the study of the evolution of the forebrain. *Journal of Comparative Neurology, 56*: 337–381.

Jones, C. and Galison, P. L. (1998). *Picturing Science, Producing Art*. New York: Routledge.

Joseph, C. M. (2011). *Stravinsky's Ballets*. New Haven: Yale University Press.

Jourdain, R. (2002). *Music, the Brain and Ecstasy*. New York: Harper Collins.

Juslin, P. N. (2001). A Brunswikian approach to emotional communication in music performance. In K. R. Hammond and T. R. Stewart (eds.), *The Essential Brunswik* (pp. 426–430). Oxford: Oxford University Press.

Juslin, P. N. and Sloboda, J. A. (2001). *Music and Emotion*. Oxford: Oxford University Press.

Juslin, P. N. and Vastfjall, D. (2008). Emotional responses to music: the need to consider underlying mechanisms. *Behavioral and Brain Sciences, 3*: 559–621.

Justus, T. C. and Bharucha, J. J. (2001). Modularity in musical processing: the automaticity of harmonic priming. *Journal of Experimental Psychology: Human Perception and Performance, 27*: 1000–1011.

Kagan, J. (1984). *The Nature of the Child*. New York: Basic Books.

Kagan, J. (2002). *Surprise, Uncertainty and Mental Structure*. Cambridge, MA: Harvard University Press.

Kahneman, D., Slovic, P., and Tversky, A. (eds.) (1982). *Judgment Under Uncertainty: Heuristics and Biases*. New York: Cambridge University Press.

Kakei, S., Hoffman, D. S., and Strick, P. L. (2001). Direction of action is represented in the ventral premotor cortex. *Nature Neuroscience, 4*: 1020–1025.

Kant, I. (1787, 1965). *Critique of Pure Reason* (Translated by L. W. Beck). New York: St. Martin's Press.

Kant, I. (1792, 1951). *Critique of Judgment*. New York: Haffner Press.

Kanwal, J. S., and Ehret, G. (2006). *Behavior and Neurodynamics for Auditory Communication*. Cambridge: Cambridge University Press.

Kaplan, H. S. and Robson, A. J. (2002). The emergence of humans: The co-evolution of intelligence and longevity with intergenerational transfers. *Proceedings of the National Academy of Sciences, 99*: 10221–10226.

Keenan, J. P., Thangaraj, V., Halpern, A. R., and Schlaug, G. (2001). Absolute pitch and planum temporale. *NeuroImage, 14*: 1402–1408.

Keil, F. C. (1979). *Semantic and Conceptual Development: An Ontological Perspective*. Cambridge, MA: Harvard University Press.

Keil, F. C. (1989). *Concepts, Kinds and Cognitive Development*. Cambridge, MA: MIT Press.

Keil, F. C. (2007). Biology and beyond: domain specificity in a broader developmental context. *Human Development, 50*: 31–38.

Kelley, A. E. (1999). Neural integrative activities of nucleus accumbens subregions in relation to learning and motivation. *Psychobiology, 27*: 198–213.

Kelley, D. B. (2002). Hormonal regulation of motor output in amphibians: *Xenopus laevis* vocalizations as a model system. In D. W. Pfaff, et al. (eds.), *Hormones, Brain and Behavior* (pp. 445–468). New York: Academic Press.

Kelley, D. B. (2004). Vocal communication in frogs. *Current Opinion in Neurobiology, 14*: 751–757.

Keltner, D. and Haidt, J. (2003). Approaching awe, a moral, spiritual, and aesthetic emotion. *Cognition and Emotion, 17*: 287–314.

Kempermann, G. (2006). *Adult Neurogenesis*. Oxford: Oxford University Press.

Kennedy, M. (1964). *The Works of Ralph Vaughn Williams*. Oxford: Oxford University Press.

Kerman, J. (1972). *Listen*. Berkeley: University of California Press.

Keverne, E. B. (2004). Understanding well-being in the evolutionary context of brain development. *Proceedings of the Royal Society of London, 359*: 1349–1358.

Keverne, E. B. and Curley, J. P. (2004). Vasopressin, oxytocin and social behavior. *Current Opinion in Neurobiology, 14*: 777–783.

Keverne, E. B. and Curley, J. P. (2008). Epigenetics, brain evolution, and behavior, *Neuroendocrinology, 29*: 398–412.

Keynes, J. M. (1921, 1957). *A Treatise on Probability*. New York: Harper & Row.

Khalfa, S., Schon, D., Anton, J., and Liegeois-Chauvel, C. (2005). Brain regions involved in the recognition of happiness and sadness in music. *Neuroreport, 16*: 1981–1984.

Kinzler, K., Dupoux, E., and Spelke, E. S. (2007). The native language of social cognition, *Proceedings of the National Academy of Sciences of the United States, 104*: 12577–12580.

Kirnarskaya, D. (2003). *The Natural Musician: On Abilities, Giftedness and Talent*. New York: Oxford University Press.

Kirschner, S. and Tomasello, M. (2009). Joint drumming: Social context facilitates synchronization in preschool children. *Journal of Experimental Child Psychology, 102*: 299–314.

Kirschner, S. and Tomasello, M. (2010). Joint music making promotes prosocial behavior in 4-year old children. *Evolution and Human Behavior, 31*: 354–364.

Kitcher, P. and Schacht, R. (2004) *Finding an Ending on Wagner's Wing*. Oxford: Oxford University Press.

Kivy, P. (1959). Charles Darwin on music. *Journal of the American Musicological Society, 12*: 42–48.

Kivy, P. (1991). Is music an art? *Journal of Philosophy, 88*: 544–554.

Kivy, P. (2001). *New Essays on Musical Understanding*. Oxford: Oxford University Press.

Kline, M. (1959). *Mathematics and the Physical World*. New York: Dover Press.

Knoblich, G. and Sebanz, N. (2006) The social nature of perception. *Current Directions in Psychological Science, 15*: 99–104.

Knops, A., Thirion, B., Hubbard, E. M., Michel, V., and Dehaene, S. (2009). Recruitment of an area involved in eye movements during mental arithmetic. *Science, 324*: 1583–1585.

Knutson, B., Wimmer, G. E., Kuhnen, C. M., and Winkielman, P. (2008). Nucleus accumbens activation mediates the influence of reward cues on financial risk taking. *Brain Imaging, 19*: 509–513.

Kochno, B. (1960). *Diaghilev and the Ballets Russes*. New York: Harper & Row.

Koelsch, S. (2005). Neural substrates of processing syntax and semantics in music. *Current Opinion in Neurobiology, 15*: 207–212.

Koelsch, S. (2006). Significance of Broca's area and ventral premotor cortex for music-syntactic processing. *Cortex, 42*: 518–520.

Koelsch, S. (2009). A neuroscientific perspective on music therapy. *The Neurosciences and Music III–Disorders and Plasticity: Annals of the New York Academy of Science, 1169*: 374–384.

Koelsch, S. (2011). Toward a neural basis of music perception. *Frontiers in Psychology, 9*: 578–584.

Koelsch, S., Fritz, T., and Schlaug, G. (2008). Amygdala activity can be modulated by unexpected chord functions during music listening. *NeuroReport, 19*: 1815–1819.

Koelsch, S., Fuermetz, J., Sack, U., Bauer, K., Hohenadel, M., Wiegel, M., et al. (2011). Effects of music listening on cortisol levels and propofol consumption during spinal anesthesia. *Frontiers in Psychology, 2*: 1–9.

Koelsch, S., Gunter, T. C., Cramon, D. Y., Zysset, S., Lohmann, G., and Friederici, A. D. (2002). Bach speaks: A cortical "language-network" serves the processing of music. *Neuroimage, 17*: 956–966.

Koelsch, S., Gunter, T., Friederici, A. D., and Schroger, E. (2000). Brain indices of music processing: "Nonmusicians" are musical. *Journal of Cognitive Neuroscience, 12*: 520–541.

Koelsch, S., Gunter, T., Schroger, E., and Friederici, A. D. (2003). Processing tonal modulations: An ERP study. *Journal of Cognitive Neuroscience, 15*: 1149–1159.

Koelsch, S., Kasper, E., Sammler, D., Schulze, K., Gunter, T., and Friederici, A. D. (2004). Music, language and meaning: Brain signature of semantic processing. *Nature Neuroscience, 7*: 302–307.

Koelsch, S., Offermans, K., and Franzke, P. (2010). Music in the treatment of affective disorders: An exploratory investigation of a new method for music-therapeutic research. *Music Perception: An Interdisciplinary Journal, 27*: 307–316.

Koelsch, S., Schroger, E., and Gunter, T. C. (2002). Music matter: preattentive musicality of the human brain. *Psychophysiology, 39*: 38–48.

Koelsch, S., and Steinbeis, N. (2008). Shared neural resources between music and language indicate semantic processing of musical tension-resolution patterns. *Cerebral Cortex, 18*: 1169–1178.

Kohut, H. (2011). *The Search for the Self: The Selected Writings of Heinz Kohut: 1950–1978,* volume 2, Paul H. Ornstein (ed.). London: Karnac Books.

Kornblith, H. (1993). *Inductive Inference and Its Natural Ground.* Cambridge, MA: MIT Press.

Kosfeld, M., Heinrichs, M., Zak, P. J., Fischbacher, U., and Fehr, E. (2005). Oxytocin increases trust in humans. *Science, 435*: 673–676.

Kosslyn, S. M. (1986). *Image and Mind.* Cambridge, MA: Harvard University Press.

Kosslyn, S. M., Alpert, N. M., Thompson, W. L., Maljkovic, V., Weise, S. B., Chabris, C. F., et al. (1993). Visual mental imagery activates topographically organized visual cortex: PET investigations. *Journal of Cognitive Neuroscience, 5*(3): 263–287.

Kramer, L. (2002). *Musical Meaning.* Berkeley: University of California Press.

Kraus, N. and Chandrasekaran, B. (2010). Music training for the development of auditory skills. *Nature Neuroscience, 11*: 599–605.

Krause, J., Lalueza-Fox, C., Orlando, L., Enard, W., Green, R. E., Burbano, H. A., et al. (2007). The derived FOXP2 variant of modern human was shared with Neandertals. *Current Biology, 17*: 1–5.

Krieckhaus, E. E. (1970). Innate recognition aids rats in sodium regulation. *Journal of Comparative and Physiological Psychology, 73*: 117–122.

Kringelbach, M. L. and Berridge, K. C. (2010). *Pleasures of the Brain.* Oxford: Oxford University Press.

Kripke, S. (1980). *Naming and Necessity.* Cambridge, MA: Harvard University Press.

Krumhansl, C. L. (2002). Music: A link between cognition and emotion. *Current Directions in Psychological Science, 11*: 45–50.

Kruse, F. E. (2005). Emotion in musical meaning: A Peircean solution to Langer's dualism. *Transactions of the Charles S. Peirce Society, 41*(4): 762–778.

Kruse, F. E. (2007). Vital rhythm and temporal form in Langer and Dewey. *Journal of Speculative Philosophy, New Series, 21*: 16–26.

Kruse, F. E. (2011). Temporality in musical meaning: A Peircean/Deweyan semiotic approach. *The Pluralist, 6*: 50–63.

Lahav, A. Boulanger, A., Schlaug, G., and Saltzman, E. (2005). The power of listening: auditory-motor interactions in musical training. *Annals of the New York Academy of Science, 1060*: 189–194.

Lahr, M. M. and Foley, R. A. (1998). Towards a theory of modern human origins: geography, demography, and diversity in recent human evolution. *American Journal of Physical Anthropology, Supplement, 27*: 137–176.

Lai, C.S.L., Fisher, S. E., Hurst, J. A., Vargha-Khadem, F., and Monaco, A. P. (2001). A forkhead-domain gene is mutated in a severe speech and language disorder. *Nature, 413*: 519–523.

Lakoff, G. and Johnson, M. (1999). *Philosophy in the Flesh: the Embodied Mind and its Challenge to Western Thought.* New York: Basic Books.

Lakoff, G. and Numez, R. E. (2000). *Where Mathematics Comes From.* New York: Basic Books.

Lamarck, J. B. (1809/1984). *Zoological Philosophy* (Translated by H. Elliot). Chicago: University of Chicago Press.

Lambert, P. (1997). *The Music of Charles Ives*. New Haven: Yale University Press.

Landau, B. and Hoffman, J. E. (2005). Parallels between spatial cognition and spatial language: Evidence from Williams syndrome. *Journal of Memory and Language, 53*: 163–185.

Langer, S. K. (1937). *Philosophy in a New Key*. Cambridge, MA: Harvard University Press.

Langer, S. K. (1953). *Feeling and Form*. New York: Scribner.

Langer, S. K. (1957). *Problems of Art*. New York: Scribner.

Langer, S. K. (1962). *Philosophical Sketches*. New York: Mentor Books.

Langer, S. K. (1973). *Mind: An Essay on Human Feeling*. Baltimore: Johns Hopkins University Press.

Larhammar, D., Dreborg, S., Larsson, T. A., and Sundstrom, G. (2009). Early duplication of opioid receptor and peptide genes in vertebrate evolution. *Trends in Comparative Endocrinology and Neurobiology, 1163*: 451–453.

Lashley, K. S. (1938). An experimental analysis of instinctive behavior. *Psychological Review, 45*: 445–471.

Lashley, K. S. (1951). The problem of serial order in behavior. In L. A. Jeffres (ed.), *Cerebral Mechanisms in Behavior* (pp. 110–133). New York: Wiley.

Lawson, G., Scarre, C., Cross, I., and Hills, C. (1998). Mounds, megaliths, music and mind: some reflections on the acoustical properties and purposes of archeological spaces. *Archeological Review from Cambridge, 15*: 111–134.

Lazarus R. S. (1984). On the primacy of cognition. *American Psychologist, 39*: 124–129.

Leakey, L.S.B. (1958). Recent discoveries at Olduvai Gorge, Tanganyika. *Nature, 181*: 1099–1103.

Leakey, M. G., Spoor, F., Bown, F. H., Gathogo, P. N., Kiarie, C., Leakey, L. N., et al. (2001). New hominin genus from eastern Africa shows diverse middle Pliocene lineages. *Nature, 410*: 433–440.

Leaver, A. M., Van Lare, J., Zielinski, B., Halpern, A. R., and Rauschecker, J. P. (2009). Brain activation during anticipation of sound sequences. *Journal of Neuroscience, 29*: 2477–2485.

Le Doux, J. E. (1996). *The Emotional Brain*. New York: Simon & Schuster.

Le Doux, J. E. (2000). Emotion circuits in the brain. *Annual Review of Neuroscience, 23*: 155–184.

Lee, K. M., Skoe, E., Kraus, N., and Ashley, R. (2009). Selective subcortical enhancement of musical intervals in musicians. *Journal of Neuroscience, 29*: 5832–5840.

Leibniz, G. (1685/1962). *Discourse on Metaphysics* (Translated by G. Montgomery). La Salle: Open Court.

Lerdahl, F. (1992). Cognitive constraints on compositional systems. *Contemporary Music Review* 6: 97–121.

Lerdahl, F. and Jackendoff, R. (1983). *A Generative Theory of Tonal Music*. Cambridge, MA: MIT Press.

Lerdahl, F. and Jackendoff, R. (1999). *A Generative Theory of Music*. Cambridge, MA: MIT Press.

Leslie, A., Gelman, R., and Gallistel, C. R. (2008). The generative basis of natural number concepts. *Trends in Cognitive Science, 12*: 213–218.
Leung, C. H., Goode, C. T., Young, L. J., and Maney, D. L. (2009). Neural distribution of nonapeptide binding sites in two species of songbird. *Journal of Comparative Neurology, 513*: 197–208.
Levi, I. (1967). *Gambling With Truth*. Cambridge, MA: MIT Press.
Levinson, S. (2006). Cognition at the heart of human interaction. *Discourse Studies, 8*: 85–93.
Levinson, S. and Jaisson, P. (eds.) (2006). *Evolution and Culture*. Cambridge, MA: MIT Press, 105–132.
Levi-Strauss, C. (1969). *The Raw and the Cooked*. New York: Harper & Row.
Levitin, D. J. (2005). Musical behavior in a neurogenetic developmental disorder. *Annals of the New York Academy of Sciences, 1060*: 1–10.
Levitin, D. J. (2006). *This Is Your Brain in Music*. New York: Plume Press.
Levitin, D. J. and Bellugi, U. (2006). Rhythm, timbre, and hyperacusis in Williams-Beuren syndrome. In C. Morris, H. Lenhoff, and P. Wang (eds.). *Williams-Beuren Syndrome: Research and Clinical Perspectives* (pp. 343–358). Baltimore, MD: Johns Hopkins University Press.
Levitin, D. J., Cole, K., Chiles, M., Lai, Z., Lincoln, A., and Bellugi, U. (2004). Characterizing the musical phenotype in individuals with Williams syndrome. *Child Neuropsychology, 10*: 223–247.
Levitin, D. J. and Menon, V. (2003) Musical structure is processed in language areas of the brain. *Neuroimage* 20: 2142–2152.
Levitin, D. J., Menon, V., Schmitt, J. E., Eliez, S., White, C. D., and Glover, G. H., et al. (2003). Neural correlates of auditory perception in Williams syndrome: An fMRI study. *NeuroImage, 18*: 74–82.
Levman, B. (1992). The genesis of music and language. *Ethnomusicology, 36*: 147–170.
Lewis, K. P. and Barton, R. A. (2006). Amygdala size and hypothalamus predict social play frequency in non-human primates. *Journal of Comparative Psychology, 120*: 31–37.
Li, X., Jarvis, E. D., Alvarez-Borda, B., Lim, D. A., and Nottebohm, F. (2000). A relationship between behavior, neurotrophin expression and new neuron survival. *Proceedings of the National Academy of Sciences of the United States, 97*: 8584–8589.
Lidov, D. (2005). *Is Language a Music?* Bloomington: University of Indiana Press.
Lieberman, D. E. (2007). Homing in on early Homo. *Nature, 499*: 291–292.
Lieberman, D. E. (2011). *The Evolution of the Human Head*. Cambridge, MA: Harvard University Press.
Lieberman, P. (1984). *The Biology and Evolution of Language*. Cambridge, MA: Harvard University Press.
Lieberman, P. (2002). *Human Language and our Reptilian Brain*. Cambridge, MA: Harvard University Press.
Lieberman, P. (2009). FOXP2 and human cognition. *Cell, 137*: 800–802
Lieberman, P. and McCarthy, R. (2007). Tracking the evolution of language and speech. *Expedition, 49*: 15–20.
Liebert, G. (2004). *Nietzsche and Music* (Translated by D. Pellauer and G. Parkes). Chicago: University of Chicago Press.

Lim, M. M., Bielsky, I. F., and Young, L. J. (2005). Neuropeptides and the social brain. *International Journal of Developmental Neuroscience, 23*: 235–243.

Liu, W. C., Gardner, T. J., and Nottebohm, F. (2004) Juvenile zebra finches can use multiple strategies to learn the same song. *Proceedings of the National Academy of Sciences, 101*: 18177–18182.

Livingstone, S. R. and Thompson W. F. (2009). The emergence of music from the *Theory of Mind. Musicae Scientiae, Special Issue, 209*: 83–115.

Lloyd, M. (1949). *The Borzoi Book of Modern Dance*. New York: Knopf.

Loewenstein, G. F. (1994). The psychology of curiosity. *Psychological Bulletin, 116*: 75–98.

Loewenstein, G. F. (1996). Out of control: Visceral influences on behavior. *Organizational Behavior and Human Decision Processes, 35*: 272–292.

Loewenstein, G. F., and Lerner, J. (2003). The role of emotion in decision making. In R. J. Davidson, H. H. Goldsmith, and K. R. Scherer (eds.), *Handbook of Affective Science* (pp. 619–642). Oxford: Oxford University Press.

Lorenz, K. (1970, Vol. 1). *Studies in Animal and Human Behaviour*. Cambridge, MA: Harvard University Press.

Loui, P., Guenther, F. H., Mathys, C., and Schlaug, G. (2008). Action-perception mismatch in tone-deafness. *Current Biology, 18*: R331–R332.

Loui, P., Li, H. C., Hohmann, A., and Schlaug, G. (2010). Enhanced cortical connectivity in absolute pitch musicians: A model for local hyperconnectivity. *Journal of Cognitive Neuroscience, 23*: 1015–1026.

MacDowell, K. A. and Mandler, G. (1989). Constructions of emotion: discrepancy, arousal, and mood. *Motivation and Emotion, 13*: 105–124.

Maess, B., Koelsch, S, Gunter, T. C., and Friederici, A. D. (2001). Musical syntax is processed in Broca's area: an MEG study. *Nature Neuroscience, 4*: 540–545.

Maidhof, C., Vavatzanidis, N., Prinz, W., Rieger, M., and Koelsch, S. (2010). Processing expectancy violations during music performance and perception: An ERP study. *Journal of Cognitive Neuroscience, 22*: 2401–2413.

Maier, N.R.F. and Schneirla, T. C. (1935/1964). *Principles of Animal Psychology*. New York: Dover Press.

Mampe, B., Friederici, A. D., Christophe, A., and Wermke, K. (1994). Newborns' cry melody is shaped by their native language. *Current Biology, 19*: 1994–1997.

Mandler, J. M. (2004). *The Foundations of Mind*. Oxford: Oxford University Press.

Maney, D. L., Goode, C. T., and Wingfield, J. C. (1997). Intraventricular infusion of arginine vasotocin induces singing in a female songbird. *Journal of Endocrinology, 9*: 487–491.

Manzon, L. A. (2002). The role of prolactin in fish osmoregulation: A review. *General and Comparative Endocrinology, 125*: 291–310.

Marin, M. M. (2009). Effects of early musical training on musical and linguistic syntactic abilities. *Annals of the New York Academy of Sciences, 1169*: 187–190.

Marler, C. A., Chu, J., and Wilczynski, W. (1995). Arginine vasotocin injection increases probability of calling in cricket frogs, but causes call changes characteristic of less aggressive males. *Hormones and Behavior, 29*: 554–570.

Marler, C. A., Boyd, S. K., and Wilczynski, W. (1999). Forebrain arginine vasotocin correlates of alternative mating strategies in cricket frogs. *Hormones and Behavior, 36*: 53–61.

Marler, P. (1961). The logical analysis of animal communication. *Journal of Theoretical Biology, 1*: 295–317.
Marler, P. (1981). Birdsong: The acquisition of a learned motor skill. *Trends in Neuroscience, 4*: 88–94.
Marler, P. (1991). Song-learning behavior: the interface with neuroethology. *Trends in Neuroscience, 14*: 199–206.
Marler, P. (2000). Origins of music and speech Insights from animals. In N. L. Wallin, B. Merker, and S. Brown (eds.), *The Origins of Music* (pp. 31–48). Cambridge, MA: MIT Press.
Marler, P. and Doupe, A. J. (2000). Singing in the Brain. *Proceedings of the National Academy of Sciences, 97*: 2965–2967.
Marler, P. and Hamilton, W. J. (1966). *Mechanisms of Animal Behavior*. New York: Wiley.
Marler, P., Peters, S., Ball, G. F., Duffy, A. M., Jr., and Wingfield, J. C. (1988). The role of sex steroids in the acquisition and production of birdsong. *Nature, 336*: 770–772.
Marler, P. and Pickert, R. (1984). Species-universal microstructure in the learning of the Swamp Sparrow. *Animal Behavior, 32*: 673–689.
Marsden, C. D. (1984). The pathophysiology of movement disorders. *Neurological Clinics, 2*: 435–459.
Marsden, C. D. and Obeso, J. A. (1994). The functions of the basal ganglia and the paradox of stereotaxic surgery in Parkinson's disease. *Brain, 117*: 877–897.
Martin, A., Wiggs, C. L., Ungerleider, L. G., and Haxby, J. V. (1996). Neural correlates of category specific knowledge. *Nature, 379*: 649–652.
Martin, A. (1998). Organization of semantic knowledge and the origin of words in the brain. *Origin of Diversification of Language, 24*: 69–87.
Martin, A. (2007). The representation of object concepts in the brain. *Annual Review of Psychology, 58*: 25–45.
Mayhew, P. J., Jenkins, G. B., and Benton, T. G. (2008). A long term association between global temperature and biodiversity, origination and extinction in the fossil record. *Proceedings of the Royal Society of London B, 275*: 47–53.
Maynard Smith, J. (1982). *Evolution and the Theory of Games*. Cambridge, UK: Cambridge University Press.
Mayr, E. (1942/1982). *Systemics and the Origin of Species*. New York: Columbia University Press.
Mayr, E. (1963). *Animal Species and Evolution*. Cambridge, MA: Harvard University Press.
McCarthy, M. M. (2008). Estradiol and the developing brain. *Physiological Review, 88*: 91–134.
McDermott, J. (2008). The evolution of music. *Nature, 453*: 287–288.
McDermott, J. and Hauser, M. (2005). The origins of music: innateness, uniqueness, and evolution. *Music Perception, 23*: 29–59.
McDonald, C. and Stewart, L. (2008). Uses and functions of music in congenital amusia. *Music Perception, 25*: 345–355.
McDonald, I. (2006). Musical alexia with recovery: a personal account. *Brain, 129*: 2554–2561.
McEwen, B. S. (1995). Steroid actions on neuronal signaling. *Ernst Schering Research Foundation Lecture Series, 27*: 1–45.

McGaugh, J. L., Cahill, L., and Roozendaal, B. (1996). Involvement of the amygdala in memory storage: Interactions with other brain systems. *Proceedings of the National Academy of Sciences, 93*: 13508–13514.
McGaugh, J. L. (2003). *Memory and Emotion.* New York: Columbia University Press.
McHenry, H. M. (1994). Tempo and mode in human evolution. *Proceedings of the National Academy of Sciences, 91*: 6780–6786.
McHenry, H. M. (2009). Human evolution. In M. Ruse and J. Travis (eds.), *Evolution: The First Four Billion Years* (pp. 256–280). Cambridge, MA: Harvard University Press.
McIntosh, G. C., Brown, S. H., Rice, R. R., and Thaut, M. H. (1997). Rhythmic auditory-motor facilitation of gait patterns in patients with Parkinson's disease. *Journal of Neurology, Neurosurgery and Psychiatry, 62*: 22–26.
McMullen, E. and Saffran, J. R. (2004). Music and language: A developmental comparison, *Music Perception, 21*: 289–311.
Mead, G. H. (1934). *Mind, Self, and Society: From the Standpoint of a Social Behaviorist.* Chicago: University of Chicago Press.
Mellars, P. (1996). *The Neanderthal Legacy.* Princeton: Princeton University Press.
Mellars, P. (2004). Neanderthals and the modern human colonization of Europe. *Nature, 432*: 461–465.
Mellars, P. (2006a). Why did modern human populations disperse from Africa 60,000 years ago? *Proceedings of the National Academy of Sciences of the United States, 103*: 9381–9386.
Mellars, P. (2006b). Going east: New genetic and archaeological perspectives on the modern human colonization of Eurasia. *Science, 313*: 796–800.
Meller, W. (1950). *Music and Society.* New York: Roy Publishers.
Menon, V. and Levitin, D. J. (2005) The rewards of music listening: response and physiological connectivity of the mesolimbic system. *Neuroimage* 28: 175–184.
Merker, B. (2005). The conformal motive in birdsong, music, and language: An introduction. *Annals of the New York Academy of Sciences, 1060*: 17–28.
Merleau-Ponty, M. (1942/1967). *The Structure of Behavior.* Boston: Beacon Press.
Merleau-Ponty, M. (1962). *The Phenomenology of Perception.* New York: Routledge and Kegan Paul.
Mervis, J. (2010). Did working memory spark creative culture? *Science, 328*: 160–161.
Meulder, M. (2010). *Helmholtz: From Enlightenment to Neuroscience.* Cambridge, MA: MIT Press.
Meyer, L. B. (1956). *Emotion and Meaning in Music.* Chicago: University of Chicago Press.
Meyer, L. B. (1967). *Music, the Arts and Ideas.* Chicago: University of Chicago Press.
Meyer, L. B. (1973). *Explaining Music: Essays and Explorations.* Los Angeles: University of California Press.
Meyer, L. B. (2000). *The Spheres of Music: A Gathering of Essays.* Chicago: University of Chicago Press.
Miell, D., MacDonald, R., and Hargreaves, D. (eds.) (2005). *Musical Communication.* Oxford: Oxford University Press.

Miller, G. A., Galanter, E., and Pribram, E. H. (1960). *Plans and the Structure of Behavior*. New York: Holt, Rinehart and Winston.

Miller, N. E. (1957). Experiments on motivation. Studies combining psychological, physiological, and pharmacological techniques. *Science, 126*: 1271–1278.

Miller, N. E. (1959). Liberalization of basic S-R concepts: Extensions to conflict behavior, motivation and social learning. In S. Koch (ed.), *Psychology: A Study of a Science,* Vol. 2 (pp. 196–292). New York: McGraw-Hill.

Miller, R. E. and Ogawa, N. (1963). Role of facial expression in co-operative avoidance conditioning of rhesus monkeys. *Journal of Abnormal and Social Psychology, 67*: 24–30.

Milner, D. and Goodale, M. A. (1995). *The Visual Brain in Action*. Oxford: Oxford University Press.

Ming, G. L. and Song, H. (2005). Adult neurogenesis in the mammalian central nervous system. *Annual Review of Neuroscience, 28*: 223–250.

Miranda, R. A. and Ullman, M. T. (2007). Double dissociation between rules and memory in music: An event-related potential study. *NeuroImage, 38*: 331–345.

Mishkin, M. and Petri, H. L. (1984). Memories and habits: Some implications for the analysis of learning and retention. In N. Butters and L. R. Squire (eds.), *Neuropsychology of Memory* (pp. 287–296). New York: Guilford.

Mithen, S. (1996). *The Prehistory of the Mind. The Cognitive Origins of Art and Science*. London: Thames and Hudson, Ltd.

Mithen, S. (2006). *The Singing Neanderthal.* Cambridge, MA: Harvard University Press.

Mithen, S. (2009). The music instinct: The evolutionary basis of musicality. *Annals of the New York Academy of Sciences, 1169*: 3–12.

Moore, F. L. (1992). Evolutionary precedents for behavioral actions of oxytocin and vasopressin. *Annals of the New York Academy of Sciences, 652*: 156–165.

Moore, F. L. and Rose, J. D. (2002). Sensorimotor processing model. In D. W. Pfaff, et al. (eds.), *Hormones, Brain and Behavior*. New York: Academic Press.

Moore-Ede, M. C., Sulzman, F. M., and Fuller, C. A. (1982). *The Clocks That Time Us*. Cambridge, MA: Harvard University Press.

Moreno, J. D. (1995). *Deciding Together*. Oxford: Oxford University Press.

Morgan, B. (1941/1980). *Martha Graham*. Dobbs Ferry: Morgan and Morgan.

Morgan, M. A., Schulkin, J., and LeDoux, J. E. (2003). Ventral medial prefrontal cortex and emotional perseveration: The memory for prior extinction training. *Behavioral Brain Research, 146*: 121–130.

Morley, I. (2002). Evolution of the physiological and neurological capacities for music. *Cambridge Archeological Journal, 12*: 195–216.

Morley, I. (2003). *The evolutionary origins and archeology of music: and investigation into the prehistory of human musical capacities and behaviors.* Unpublished PhD diss., University of Cambridge, Cambridge.

Morley, I. (2006). Mousterian Musicianship? The case of the Divje Babe 1 bone. *Oxford Journal of Archaeology, 25*: 317–333.

Morris, C. (1938/1979). *Foundations of the Theory of Signs*. Chicago: University of Chicago Press.

Morrison, S. J., Demorest, S. M., and Stambaugh, L. A. (2008). Enculturation effects in music cognition: The role of age and music complexity. *Journal of Research in Music Education, 56*: 118–129.

Morton, E. S. (1977). On the occurrence and significance of motivation-structural rules in some bird and mammal sounds. *American Naturalist, 111*: 855–869.

Musacchia, G., Sams, M., Skoe, E., and Kraus, N. (2007). Musicians have enhanced subcortical auditory and audiovisual processing of speech and music. *Proceedings of the National Academy of Sciences, 104*: 15894–15898.

Myers, C. S. (1905). A study of rhythm in primitive peoples. *British Journal of Psychology, 1*: 397–405.

Myers, C. S. (1913). The beginnings of music. In E. Quiggin (ed.), *Essays and Studies Presented to William Ridgeway.* Cambridge: Cambridge University Press.

Narmour, E. (1990) *The Analysis and Cognition of Basic Melodic Structures.* Chicago: University of Chicago Press.

Narmour, E. (2008). My intellectual father. *Music Perception, 25*: 485–487.

Nauta, W.J.H. (1972). The central visceromotor system: a general survey. In C. H. Hockman (ed.), *Limbic System Mechanisms and Autonomic Function.* Springfield: Charles C. Thomas.

Nauta, W.J.H. and Feirtag, M. (1986). *Fundamental Neuroanatomy.* San Francisco: Freeman.

Nelson, R. J., Demas, G. E., Klein, S. L., Kriegsfeld, L. J., and Bronson, F. (2002). *Seasonal Patterns of Stress Immune Function and Disease.* Cambridge: Cambridge University Press.

Nephew, B. C. and Bridges, R. S. (2008). Arginine vasopressin V1a receptor antagonist impairs maternal memory in rats. *Physiology and Behavior, 95*: 182–186.

Nettl, B. (1948). *The Book of Musical Documents.* New York: Philosophical Library.

Nettl, B. (1956) *Music in Primitive Culture.* Cambridge, MA: Harvard University Press.

Nettl, B. (2005). *The Study of Ethnomusicology: Thirty-One Issues and Concepts* (2nd ed.). Urbana: University of Illinois Press.

Nettl, B. (2006). Response to Victor Grauer: On the concept of evolution in the history of ethnomusicology. *The World of Music, 48*: 59–72.

Neuhaus, C., Knosche, T. R., and Friederici, A. D. (2006). Effects of musical expertise and boundary markers on phrase perception in music. *Journal of Cognitive Neuroscience, 18*: 472–493.

Neville, R. C. (1974). *The Cosmology of Freedom.* New Haven: Yale University Press.

Newman, S. W. (2002). Phermonal signals access the medial extended amygdala: One node in a proposed social behavior network. In D. W. Pfaff, et al. (eds.), *Hormones, Brain and Behavior* (pp. 17–32). New York: Academic Press.

Nice, M. M. (1943) Studies in the life history of the song sparrow. *Transactions of the Linnean Society of London, 6*: 1–32.

Nichols, R. (2011). *Ravel.* New Haven: Yale University Press.

Nietzsche, F. (1871/1927). *The Birth of Tragedy from the Spirit of Music.* New York: Modern Library.

Noe, A. (2004). *Action in Perception.* Cambridge, MA: MIT Press.

Noonan, J. P., Coop, G., Judaravalli, S., Smith, D., Krause, J., Alessi, J., et al. (2006). Sequencing and analysis of Neanderthal genomic DNA. *Science, 314*: 1113–1118.

Norgren, R. (1995). Gustatory system. In *The Rat Nervous System* (2nd ed.). San Diego: Academic Press, pp. 751–771.

Norton, A., Winner, E., Cronin, K., Overy, K., Lee, D. J., and Schlaug, G. (2005) Are there pre-existing neural, cognitive or motoric markers for musical ability? *Brain and Cognition, 59*: 124–134.

Nottebohm, F. (1970). Ontogeny of bird song. *Science, 167*: 950–956.

Nottebohm, F. (1994). The song circuits of the avian brain as a model system in which to study vocal learning, communication and manipulation. *Discussions in Neurosciences, 10*: 72–81.

Nottebohm, F. (2005). The neural basis of birdsong. *PLOS Biology, 3*: 164–172.

Nottebohm, F. (2008). Auditory experience and song development in the chaffinch Fringilla Coelebs. *International Journal of Avian Science, 110*: 549–568.

Nussbaum, M. C. (2001). *Upheavals of Thought: The Intelligence of Emotion.* Cambridge, UK: Cambridge University Press.

O'Connor, R. C. (2007). Dolphin social intelligence: Complex alliance relationships in bottlenose dolphins and a consideration of selective environments for extreme brain size evolution in mammals. *Philosophical Transactions of the Royal Society of London B, Biological Sciences, 362*: 587–602.

O'Doherty, J., Dayan, P., Schultz, J., Deichmann, R., Friston, K., and Dolan, R. J. (2004). Dissociable roles of ventral and dorsal striatum in instrumental conditioning. *Science, 304*: 452–455.

Oftedal, O. T. (2002). The origin of lactation as a water source for parchment shelled eggs. *Journal of Mammary Gland Biology and Neoplasia, 7*: 253–266.

Özdemir, E. (2006). Neural correlates of singing and speaking. *NeuroImage, 67*: 2854.

Özdemir, E., Norton, A., and Schlaug, G. (2006). Shared and distinct neural correlates of singing and speaking. *NeuroImage, 33*: 628–635.

Palmer, C. and Jungers, M. K. (2003). Music cognition. In L. Nadel (ed.), *Encyclopedia of Cognitive Science* (pp. 155–158). New York: Wiley.

Palmer, C., Jungers, M. K., and Jusczyk, P. W. (2001). Episodic memory for musical prosody. *Journal of Memory and Language, 45*: 526–545.

Palmer, C., Kopmans, E., Loehr, J. D., and Carter, C. (2009). Movement-related feedback and temporal accuracy in clarinet performance. *Music Perception, 26*: 439–450.

Panksepp, J. (1995). The emotional sources of "chills" induced by music. *Music Perception, 13*: 171–207.

Panksepp, J. (1998). *Affective Neuroscience: The Foundations of Human and Animal Emotions.* New York: Oxford University Press.

Panksepp, J. and Bernatzky, G. (2002). Emotional sounds and the brain: The neuro-affective foundations of musical appreciation. *Behavioural Processes, 60*: 133–155.

Parkinson, J. A., Crofts, H. S., McGuigan, M., Tomic, D. L., Everitt, B. J., and Roberts, A. C. (2001). The role of the primate amygdala in conditioned reinforcement. *Journal of Neuroscience, 21*: 7770–7780.

Parrott, G. W. and Schulkin, J. (1993). Neuropsychology and the cognitive nature of emotions. *Cognition and Emotion, 7*: 43–59.

Pashler, H. E. (1998). *The Psychology of Attention.* Cambridge, MA: MIT Press.

Passingham, R. (2008). *What Is Special about the Human Brain*. Oxford: Oxford University Press.
Patel, A. D. (1998). Syntactic processing in language and music: Different cognitive operations, similar neural resources. *Music Perception, 16*: 27–42.
Patel, A. D., Gobson, E., Ratner, J., Besson, M., and Holcomb, P. J. (1998). Processing syntactic relations in language and music: An event-related potential study. *Journal of Cognitive Neuroscience, 10*: 717–733.
Patel, A. D. (2003). Language, music, syntax and the brain. *Nature Neuroscience, 6*: 674–681.
Patel, A. D. (2005). The relationship of music to the melody of speech and to syntactic processing disorders in aphasia. *Annals of the New York Academy of Sciences, 1060*: 59–70.
Patel, A. D. (2008a). *Music, Language, and the Brain*. New York: Oxford University Press.
Patel, A. D. (2008b). Talk of the tone. *Nature, 453*: 726–727.
Patel, A. D. (2010). *Music, Biological Evolution, and the Brain*. New York: Oxford University Press.
Patel, A. D., Iversen, J. R., Wassenaar, M., and Hagoort, P. (2008). Musical syntactic processing in agrammatic broca's aphasia. *Aphasiology, 22*: 776–789.
Payne, R. S. and McVay, S. (1971). Songs of humpback whales. *Science, 173*: 585–597.
Peacocke, C. (2009). The perception of music: sources of significance. *British Journal of Aesthetics, 49*: 257–275.
Peery, J. C., Peery, I. C., and Draper, T. W. (1987). *Music and Child Development*. New York: Springer-Verlag.
Peirce, C. S. (1868). Questions concerning certain faculties claimed for man. *Journal of Speculative Philosophy, 2*: 103–114.
Peirce, C. S. (1877). The fixation of belief. *Popular Science Monthly, 12*: 1–15. In *The Essential Peirce*: Vol. 1, p. 115.
Peirce, C. S. (1878). Deduction, induction and hypothesis. *Popular Science Monthly, 13*: 470–482.
Peirce, C. S. (1880). Logic, chapter 1. In *Writings of C. S. Peirce,* Vol. 4. Bloomington: Indiana University Press.
Peirce, C. S. (1899/1992). *Reasoning and the Logic of Things*. K. L. Ketner and H. Putnam (eds.), Cambridge, MA: Harvard University Press.
Peirce, C. S. (1903–1912/1977). *Semiotic and Significs: The Correspondence between Charles S. Peirce and Victoria Lady Welby*. C. S. Hardwick (ed.) Bloomington: Indiana University Press.
Peirce, C. S. (2000–2009). *Writings of C. S. Peirce,* Vols. 1–8. Bloomington: Indiana University Press.
Penhune, V. B. (2011). Sensitive periods in human development: Evidence from musical training. *Cortex, 47*: 1126–1137.
Penhune, V., Watanabe, D., and Savion-Lemieux, T. (2005) The effect of early musical training on adult motor performance. *Annals of the New York Academy of Sciences, 1060*: 265–268.
Perani, D., Cappa, S. F., Bettinardi, V., Bressi, S., Gorno-Tempini, M., Matarrese, M., et al. (1995). Different neural systems for the recognition of animals and made tools. *NeuroReport, 6*: 1636–1641.

Perani, D. Saccuman, M. C., Scifo, P., Spada, D., Andreolli, G., Rovelli, R., et al. (2010). Functional specializations for music processing in the human newborn brain. *Proceedings of the National Academy of Sciences of the United States, 107*: 4758–4763.

Percy, W. (1954/1975). *The Message in the Bottle*. New York: Noonday Press.

Peretz, I. (2006). The nature of music from a biological perspective. *Cognition, 100*: 1–32.

Peretz, I., (2008). Musical disorders: From behavior to genes. *Current Directions in Psychological Science, 17*: 329–333.

Peretz, I., Gosselin, N., Belin, P., Zatorre, R. J., Plailly, J., and Tillman, B. (2009). Music lexical networks, the cortical organization of music recognition. *The Neurosciences and Music III–Disorders and Plasticity: Annals of the New York Academy of Sciences, 1169*: 256–265.

Peretz, I. and Zatorre, R. (2003). *The Cognitive Neuroscience of Music*. Oxford: Oxford University Press.

Perlovsky, L. (2010). Musical emotions: Functions, origins, evolution. *Physics of Life Reviews, 7*: 2–27.

Perrett, D. I. and Emery, N. J. (1994). Understanding the intentions of others from visual signals: Neurophysiological evidence. *Cahiers de Pscyhologie Cognitive, 13*: 683–694.

Perrett, D. I., Hietanen, J. K., Oram, M. W., and Benson, P. J. (1992). Organization and functions of cells responsive to faces in the temporal cortex. *Philosophical Transactions of the Royal Society of London B, 335*: 23–30.

Petrovic, P., Kalisch, R., Singer, T., and Dolan, R. J. (2008). Oxytocin attenuates affective evaluations of conditioned faces and amygdala activity. *Journal of Neuroscience, 28*: 6607–6615.

Pfordresher, P. Q. (2003). The role of melodic and rhythmic accents in musical structure. *Music Perception, 20*: 431–464.

Pfordresher, P. Q. (2006). Coordination of perception and action in music performance. *Advances in Cognitive Psychology, 2*: 183–198.

Pfordresher, P. Q. and Brown, S. (2009). Enhanced production and perception of musical pitch in tone language speakers. *Attention, Perception and Psychophysics, 71*: 1385–1398.

Phoenix, C. H., Goy, R. W., Gerall, A. A., and Young, W. C. (1959). Organizing action of prenatally administered testosterone proportionate on the tissues mediating mating in the female guinea pig. *Endocrinology, 65*: 364–389.

Piaget, J. (1954). *The Construction of Reality in the Child*. New York: Basic Books.

Pinker, S. (1994). *The Language Instinct*. New York: William Morrow.

Pinker, S. and Jackendoff, R. (2005). The faculty of language: What's special about it? *Cognition, 85*: 201–236.

Pitkow, L. J., Sharer, C. A., Ren, X., Insel., T. R., Terwilliger, E. F., and Young, L. J. (2001). Facilitation of affiliation and pair-bond formation by vasopressin receptor gene transfer into the ventral forebrain of a monogamous vole. *Journal of Neurosciences, 21*: 7392–7396.

Plato. (1985). *Meno*. Oxford: Aris and Phillips.

Porges, S. W. (2004). The vagus: A mediator of behavioral and visceral features associated with autism. In M. L. Bauman and T. L. Kemper (eds.), *The Neu-*

robiology of Autism (pp. 65–78). Baltimore, MD: Johns Hopkins University Press.

Porges, S. W. (2007). The polyvagal perspective. *Biological Psychology, 74*: 116–143.

Posner, M. I. (1990). *Foundations of Cognitive Science*. Cambridge, MA: MIT Press.

Power, M. L. and Schulkin, J. (2005). *Birth Distress and Disease*. Cambridge: Cambridge University Press.

Power, M. L. and Schulkin, J. (2009). *Evolution of Obesity*. Baltimore, MD: Johns Hopkins University Press.

Premack, D. and Premack, A. J. (1995). Origins of human social competence. In M. S. Gazzaniga (ed.), *The Cognitive Neurosciences* (pp. 205–218). Cambridge, MA: MIT Press.

Premack, D. and Woodruff, G. (1978). Does the chimpanzee have a theory of mind? *Behavioral and Brain Sciences, 1*: 515–526.

Premack, D. (1990). The infant's theory of self-propelled objects. *Cognition, 36*: 1–16.

Price, J. L. (1999). Prefrontal cortical networks related to visceral function and mood. *Annals of the New York Academy of Sciences, 877*: 383–396.

Price, J. L., Carmichael, S. T., and Drevets, W. C. (1996). Networks related to the orbital and medial prefrontal cortex: A substrate for emotional behavior? *Progress in Brain Research, 107*: 523–536.

Prinz, J. L. and Barsalou, L.W. (2000). Steering a course for embodied representation. In E. Dietrich and Markman, A. B. (eds.), *Cognitive Dynamics: Conceptual Change in Humans and Machines* (pp. 51–77). Cambridge, MA: MIT Press.

Pulvermuller, F. (2002). *The Neuroscience of Language*. Oxford: Oxford University Press.

Pulvermuller, F. Hauk, O. Nikulin, V. V., and Ilmoniemi, R. J. (2005). Functional links between motor and language systems. *European Journal of Neuroscience, 21*: 793–797.

Pulvermuller, F., Shtyrov, Y., and Ilmoniemi, R. (2005). Brain signatures of meaning in action word recognition. *Journal of Cognitive Neuroscience, 6*: 884–892.

Quine, W.V.O. (1969). Epistemology naturalized. In *Ontological Relativity and Other Essays*. New York: Columbia University Press.

Raffman, D. (1993). *Language, music, and mind*. Cambridge, MA: MIT Press.

Rakic, P. (1988). Defects of neuronal migration and pathogenesis of cortical malformations. *Progress in Brain Research, 73*: 15–37.

Rakic. P. (2002). Evolving concepts of cortical radial and areal specification. *Progress in Brain Research, 136*: 265–280.

Ramachandran, V. S. and Hirsetin, W. (1999). The science of art: a neurological theory of aesthetic experience. *Journal of Consciousness Studies, 6*: 15–51.

Rameau. J. K. (1722, 1971). *Treatise on Harmony*. New York: Dover Press.

Rathelot, J. A. and Strick, P. L. (2009). Subdivisions of primary motor cortex based on cortico-motoneuronal cells. *Proceedings of the National Academy of Sciences, 106*: 918–923.

Rauschecker, J. P. (2006). Cortical plasticity and music. *Annals of the New York Academy of Sciences,* 930: 330–336.

Rauschecker, J. P. and Korte, M. (1993). Auditory compensation for early blindness in cat cerebral cortex. *Journal of Neuroscience, 139*: 4538–4548.

Rauschecker, J. P. and Scott, S. K. (2009). Maps and streams in the auditory cortex: nonhuman primates illuminate human speech processing. *Nature Neuroscience, 12*: 718–724.

Reader, S. M. and Laland, K. N. (2002). Social intelligence, innovation, and enhanced brain size in primates. *Proceedings of the National Academy of Sciences, 99*: 4436–4441.

Reich, S. (1974). *Writing about Music.* New York: New York University Press.

Reiss, J. E., Hoffman, J. E., and Landau, B. (2005). Motion processing specialization in Williams syndrome. *Vision Research, 45*: 3379–3390.

Repp, B. H. (2001). Processes underlying adaptation to tempo changes in sensorimotor synchronization. *Human Movement Science, 20*: 277–312.

Repp, B. H. (2005). Sensorimotor synchronization: A review of the tapping literature. *Psychonomic Bulletin and Review, 12*: 969–992.

Repp, B. H. (2006). Musical Synchronization. In E. Altenmuller, M. Wiesendanger, and J. Kesserling (eds.), *Music, Motor Control, and the Brain* (pp. 55–76). Oxford: Oxford University Press.

Rescorla, R. A. (1988). Pavlovian conditioning. *American Psychologist, 43*: 151–160.

Rescorla, R. A. and Wagner, A. R. (1972). A theory of Pavlovian conditioning. In A. H. Black and W. F. Prokasy (eds.). *Classical Conditioning II: Current research and theory* (pp. 64–99). New York: Appleton Century Crofts.

Richard, J. M. and Berridge, K. C. (2011). Metabotropic glutamate receptor blockade in nucleus accumbens shell shifts affective valence towards fear and disgust. *European Journal of Neuroscience, 33*: 736–747.

Richardson, A. (2010). *The Neural Sublime.* Baltimore: Johns Hopkins University Press.

Richter, C. P. (1943). *Total Self-Regulatory Functions in Animals and Man.* New York: Harvey Lecture Series.

Richter, C. P. (1965/1979). *Biological Clocks in Medicine and Psychiatry.* Springfield: Charles C. Thomas.

Rickard, N. S. (2009). Defining the rhythmicity of memory-enhancing acoustic stimuli in the young domestic chick. *Journal of Comparative Psychology, 123*(2): 217–221.

Rimmele, U., Hediger, K., Heinrichs, M., and Klaver, P. (2009). Oxytocin makes a face in memory familiar. *Journal of Neuroscience, 29*: 38–42.

Rizzolatti, G. and Arbib, M. A. (1998). Language within our grasp. *Trends in Neuroscience, 21*: 188–194.

Rizzolatti, G. and Luppino, G. (2001). The cortical motor system. *Neuron, 31*: 889–901.

Robson, S. and Wood, B. (2008). Hominin life history: Reconstruction and evolution. *Journal of Anatomy, 212*: 394–425.

Rodrigues, S. M., Saslow, L. R., Garcia, N., John, O. P., and Keltner, D. (2009). Oxytocin receptor genetic variation relates to empathy and stress reactivity in humans. *Proceedings of the National Academy of Sciences of the United States, 106*: 21437–21441.

Rolls, E. T. (2000). The orbitofrontal cortex and reward. *Cerebral Cortex, 10*: 284–294.

Rolls, E. T. and Treves, A. (1998). *Neural Networks and Brain Function.* New York: Oxford University Press.

Rolls, E. T., Treves, A., and Tovee, M. T. (1997). The representational capacity of the distributed encoding of information provided by populations of neurons in primate visual cortex. *Experimental Brain Research, 114:* 149–162.

Ronchi, V. (1957/1991). *Optics: A Science of Vision* (Translated by E. Rosen). New York: Dover.

Rosati, A. G., Stevens, J. R., Hare, B., and Hauser, M. D. (2007). The evolutionary origins of human patience: temporal preferences in chimpanzees, bonobos and human adults. *Current Biology, 17*: 1663–1668.

Rose, J. D., Moore, F. L., and Orchinik, M. (1993). Rapid neurophysiological effects of corticosterone on medullary. *Neuroendocrinology, 57*: 815–824.

Rosen, C. (2010). *Music and Sentiment.* New Haven: Yale University Press.

Rosen, J. B. and Schulkin, J. (1998). From normal fear to pathological anxiety. *Psychological Review, 105*: 325–350.

Rosenzwieg, M. R. (1984). Experience, memory and the brain. *American Psychologist, 39*: 365–375.

Rosner, B. S. and Meyer, L. B. (1986). The perceptual roles of melodic process, contour and form. *Music Perception, 4*: 1–40.

Rosner, B. S. and Narmour, E. (1992). Harmonic closure. *Music Perception, 9*: 383–411.

Roth, G. and Dicke, U. (2005). Evolution of the brain and intelligence. *TRENDS in Cognitive Sciences, 9*(5): 250–257.

Rousseau, J. J. (1966) *On the Origins of Language*. Chicago: University of Chicago Press.

Rozin, P. (1976). The evolution of intelligence and access to the cognitive unconscious. In J. Sprague and A. N. Epstein (eds.), *Progress in Psychobiology and Physiological Psychology* (pp. 245–281). New York: Academic Press.

Rozin, P. (1998). Evolution and development of brains and cultures: Some basic principles and interactions. In M. S. Gazzaniga and J. S. Altman (eds.), *Brain and Mind: Evolutionary Perspectives*. Strassbourg: Human Frontiers Science Program.

Rozin, P., Rozin, A., Appel, B., and Wachtel, C. (2006). Documenting and explaining the common AAB pattern in music and humor: Establishing and breaking expectations. *Emotion, 6*: 349–355.

Ruiz, M. H., Jabusch, H. C., and Altenmuller, E. (2009). Detecting wrong notes in advance: Neuronal correlates of error monitoring in pianists. *Cerebral Cortex, 19*: 2625–2639.

Rumph, S. (2011). *Mozart and the Enlightenment Semiotics*. Berkeley: University of California Press.

Russo, F. A. and Cuddy, L. L. (1999). Motor theory of melodic expectancy. In *Acoustical society of America ASA/EAA/DAGA '99 meeting lay language papers*. Berlin.

Ryle, G. (1949/1990). *The Concept of Mind*. New York: Harper & Row.

Sabini, J. and Schulkin, J. (1994). Biological realism and social constructivism. *Journal for the Theory of Social Behavior, 224*: 207–217.

Sachs, C. (1937/1963). *World History of Dance*. New York: Norton.

Sachs, C. (1945). *The History of Musical Instruments*. New York: Norton Press.

Sachs, H. (2010). *The Ninth: Beethoven and the World: 1824*. New York: Farber and Farber.

Sacks, O. (2010). *The Mind's Eye*. New York: Knopf.

Sacks, O. (2008). *Musicophilia*. New York: Vintage Press.

Salimpoor, V. N., Benovoy, M., Larcher, K., Dagher, A., and Zatorre, R. J. (2011). Anatomically distinct dopamine release during anticipation and experience of peak emotion to music. *Nature Neuroscience, 14*: 257–260.

Sapoznik, H. (1999). *Klezmer: Jewish Music from Old World to Our New World*. New York: Schiermer Trade.

Satoh, M., Takeda, K., Nagata, K., Hatazawa, J., and Kuzuhara, S. (2001). Activated brain regions in musicians during an ensemble: A PET study. *Cognitive Brain Research, 12*: 101–108.

Savaskan, E., Ehrhardt, R., Schulz, A., Walter, M., and Schächinger, H. (2008). Post-learning intranasal oxytocin modulates human memory for facial identity. *Psychoneuroendocrinology, 33*: 368–374.

Saxe, R., Tzelnic, T., and Carey, S. (2006). Five-month old infants know humans are solid, like inanimate objects. *Cognition, 101*: B1–B8.

Schacter, D. L. and Tulving, E. (1994). *Memory Systems*. Cambridge, MA: MIT Press.

Scheijen, S. (2010). *Diaghliev: A Life*. Oxford: Oxford University Press.

Schellenberg, G. E., Adachi, M., Purdy, K. T., and McKinnon, M. C. (2002). Expectancy in Melody: Tests of Children and Adults. *Journal of Experimental Psychology, 131*: 511–537.

Schellenberg, G. E. and Peretz, I. (2008). Music, language and cognition: unresolved issues. *Trends in Cognitive Sciences, 12*: 45–46.

Scherer, K. R. (1995). Expression of emotion in voice and music. *Journal of Voice, 9*: 235–248.

Schiller, F. (1795/1980). *On the Aesthetic Education of Man*. New York: Frederick Ungar.

Schlaug, G. (2006). The brain of musicians. *Annals New York Academy of Sciences, 930*: 281–299.

Schlaug, G., Jancke, L., Huang, Y., Staiger, J. F., and Steinmetz, H. (1995). Increased corpus callosum size in musicians. *Neuropsychologia, 33*: 1047–1055.

Schmidt, L. A and Trainor, L. J. (2001). Frontal brain electrical activity (EEG) distinguishes valence and intensity of musical emotions. *Cognition and Emotion, 15*: 487–500.

Schneider, P., Scherg, M., Dosch, H. G., Specht, H. J., Gutschalk, A., and Rupp, A. (2002). Morphology of Heschl's gyrus reflects enhanced activation in the auditory cortex of musicians. *Nature Neuroscience, 5*: 688–694.

Schneider, P., Sluming, V., Roberts, N., Scherg, M., Goebel, R., Specht, H. J., et al. (2005). Structural and functional asymmetry of lateral Heschl's gyrus reflects pitch perception preference. *Nature Neuroscience, 8*: 1241–1247.

Schoenberg, A. (1975). *Style and Idea: Selected Writings*. New York: St. Martin's Press.

Schon, D. Gordon, R. L., and Besson, M. (2005). Musical and linguistic processing in song perception. *Annals of the New York Academy of Sciences, 1060*: 71–81.

Schonberg, T., Daw, N. D., Joel, D., and O'Doherty, J. P. (2007). Reinforcement learning signals in the human striatum distinguish learners from non-learners during reward-based decision making. *Journal of Neuroscience, 27*: 12860–12867.

Schopenhauer, A. (1844/1966, Vol. II). *The World as Will and Representation.* New York: Dover Press.

Schulkin, J. (1999). *The Neuroendocrine Regulation of Behavior.* Cambridge: Cambridge University Press.

Schulkin, J. (2000). *Roots of Social Sensibility and Neural Function.* Cambridge, MA: MIT Press.

Schulkin, J. (2003). *Rethinking Homeostasis.* Cambridge, MA: MIT Press.

Schulkin, J. (2004). *Bodily Sensibility: Intelligent Action.* Oxford: Oxford University Press.

Schulkin, J. (2006). *Effort: A Behavioral Neuroscience Perspective on the Will.* Mahwah: Lawrence Erlbaum.

Schulkin, J. (2009). *Cognitive Adaptation: A Pragmatist Perspective.* Cambridge: Cambridge University Press.

Schultz, W. (2002). Getting formal with dopamine and reward. *Neuron, 36*: 241–263.

Schultz, W. (2007). Multiple dopamine functions at different time courses. *Annual Review of Neuroscience, 30*: 59–88.

Schutz, A. (1932/1967). *The Phenomenology of the Social World* (Trans. G. Walsh and F. Lehnert). Chicago: Northwestern University Press.

Schutz, A. (1967). Making Music Together. In *Collected Works Vol. II.* The Hague: Martinus Nijhoff.

Schwabl, H. (1993). Yolk is a source of maternal testosterone for developing birds. *Annals of the New York Academy of Sciences, 90*: 11446–11450.

Schwartz, C. E., Wright, C. I., Shin, L. M., Kagan, J., Whalen, P. J., McMullin, K. G., et al. (2003). Differential amygdalar response to novel versus newly familiar neutral faces: a functional MRI probe developed for studying inhibited temperament. *Biological Psychology, 53*: 854–862.

Schwartz, M., Keller, P. E., Patel, A. D., and Kotz, S. A. (2011). The impact of the basal ganglia lesions on sensori-motor synchronization, spontaneous motor tempo and the detection of tempo changes. *Behavioral Brain Research, 216*: 685–691.

Sebeok, T. A. (2001). *Global Semiotics.* Bloomington: Indiana University Press.

Sergent, J. (1993). Music, the brain and Ravel. *Trends in Neuroscience, 16*: 106–109.

Sergent, J., Zuck, E., Terriah, S., and MacDonald, B. (1992). Distributed neural network underlying musical sight-reading and keyboard performance. *Science, 257*: 106–108.

Sevdalis, V. and Keller, P. E. (2011). Captured by motion: dance, action understanding and social cognition. *Brain and Cognition, 77*: 231–236.

Seyfarth, R. M., Cheney, D. L., and Bergman T. J. (2005). Primate social cognition and the origins of language. *Trends in Cognitive Sciences, 9*: 264–266.

Shepard, R. (1994). *The Mesh Between Mind and World.* Cambridge, MA: Harvard University Press.

Shepard, R. N. and Cooper, L. A. (1982). *Mental Images and Their Transformations.* Cambridge, MA: MIT Press.

Shettleworth, S. J. (1998). *Cognition, Evolution and Behavior.* Oxford: Oxford University Press.

Shors, T. J., Micseages, C., Beylin, A., Zhao, M., Rydel, T., and Gould, E. (2001). Neurogenesis in the adult is involved in the formation of trace memories. *Nature, 410*: 372–375.

Short, T. L. (2007). *Peirce's Theory of Signs.* Cambridge, UK: Cambridge University Press.

Shweder, R. A. (1991). *Thinking Through Cultures.* Cambridge, MA: Harvard University Press.

Siegel, M. B. (1979). *The Shapes of Change.* New York: Avon Books.

Silk, J. B. (2007). The adaptive value of sociality in mammalian groups. *Proceedings of the Royal Society of London, 362*: 539–559.

Simao, S. M. and Moreira, S. C. (2005). Vocalizations of a female humpback whale in Arraial do Cabo. *Marine Mammal Science; 21*: 150–153.

Simmons, W. K. and Martin, A. (2011). Spontaneous resting-state BOLD fluctuations reveal persistent domain-specific neural networks. *Social Cognitive and Affective Neuroscience. 7*: 467–475.

Simon, H. A. (1982). *Models of Bounded Rationality.* Cambridge, MA: MIT Press.

Simpson, G. G. (1949). *The Meaning of Evolution.* New Haven: Yale University Press.

Singer, T., Snozzi, R., Bird, G., Petrovic, P., Silani, G., Heinrichs, M., et al. (2008). Effects of oxytocin and prosocial behavior on brain responses to direct and vicariously experienced pain. *Emotion, 8*: 781–791.

Sloboda, J. A. (1985/2000). *The Musical Mind.* Oxford: Oxford University Press.

Sloboda, J. A. (1991). Music structure and emotional response: Some empirical findings. *Psychology of Music, 19*: 110–120.

Sloboda, J. A. (2000). *Generative Processes in Music.* Oxford: Clarendon Press.

Sloboda, J. A. (2005). *Exploring the Musical Mind: Cognition. Emotion, Ability, Function.* Oxford: Oxford University Press.

Sloboda, J. (2008). Embracing Uncertainty. *Music Perception: An Interdisciplinary Journal, 25*: 489–491.

Sluming, V., Barrick, T., Howard, M., Cezayirli, E., Mayes, A., and Roberts, N. (2002). Voxel-based morphometry reveals increased gray matter density in Broca's area in male symphony orchestra musicians. *NeuroImage, 17*: 1613–1622.

Sluming, V., Brooks, J., Howard, M., Downes, J. J., and Roberts, N. (2007). Broca's area supports enhanced visuospatial cognition in orchestral musicians. *Journal of Neuroscience, 27*: 3799–3806.

Smith, D. J., Kemler, Nelson, D. G., Grohskopf, L. A., and Appleton, T. (1994). What child is this? What interval was that? Familiar tunes and music perception in novice listeners. *Cognition, 52*: 23–54.

Smith, D. J. and Melara, R. J. (1990). Aesthetic preference and syntactic prototypicality in music: 'Tis the gift to be simple. *Cognition, 34*: 279–298.

Smith, D. J., Shields, W. E., Schull, J., and Washburn, D. A. (1997). The uncertain response in humans and animals. *Cognition, 62*: 75–97.

Smith, D. J. and Washburn, D. A. (2005). Uncertainty monitoring and metacognition by animals. *Current Directions in Psychological Science, 14*: 19–24.

Smith, D. J. and Witt, J. N. (1989). Spun steel and stardust: The rejection of contemporary compositions. *Music Perception, 7*: 169–186.

Smith, G. P. (1997). *Satiation From Gut to Brain.* Oxford: Oxford University Press.

Smith, J. D. (1987). Conflicting aesthetic ideals in a musical culture. *Music Perception, 4*: 373–392.

Smith, K. S., Berridge, K. C., and Aldridge, J. W. (2011). Disentangling pleasure from incentive salience and learning signals in brain reward circuitry. *Proceedings of the National Academy of Sciences of the United States, 108*: E255–E264.

Smith, J. E. (1978). *Purpose and Thought*. New Haven: Yale University Press.

Smith, J. N., Goldizen, A. W., Dunlop, R. A., and Noad, M. J. (2008). Songs of male humpback whales, Megaptera novaeangliae, are involved in intersexual interactions. *Animal Behavior, 76*: 467–477, Part 2.

Smith, W. J. (1977). *The Behavior of Communicating: An Ethological Approach*. Cambridge, MA: Harvard University Press.

Soto, D., Funes, M. J., Guzman-Garcia, A., Warbrick, T., Rothstein, P., and Humphreys, G. W. (2009). Pleasant music overcomes the loss of awareness in patients with visual neglect. *Proceedings of the National Academy of Sciences of the United States, 106*: 6011–6016.

Spelke, E. (1990). Principles of object perception. *Cognitive Science, 14*: 29–56.

Spencer, H. (1852). *Essays: Scientific, political, & speculative. Volume I*. Edinburgh: Williams and Norgate.

Spencer, H. (1855). *The Principles of Psychology*. London: Longman, Brown, Green and Longmans.

Spencer, H. (1857, 1951). The origin and function of music. In *Literary Style and Music*. New York: Philosophical Library.

Spitzer, M. (2004). *Metaphor and Musical Thought*. Chicago: University of Chicago Press.

Spoor, F., Leakey, M. G., Gathogo, P. N., Brown, F. H., Anton, S. C., McDougall, I., et al. (2007). Implications of new early Homo fossils from Ileret, east of Lake Tukana, Kenya. *Nature, 448*: 688–691.

Squire, L.R. (1987). *Memory and Brain*. New York: Oxford University Press.

Squire, L. R. (2004). Memory systems of the brain: A brief history and current perspective. *Neurobiology of Learning and Memory, 82*: 171–177.

Squire, L. R., Knowlton, B., and Musen, G. (1993). The structure and organization of memory. *Annual Review of Psychology, 44*: 453–495.

Squire, L. R. and Zola, S. M. (1996). Structure and function of declarative and non-declarative memory systems. *Annals of the National Academy of Sciences, 93*: 13515–13522.

Sridharan, D., Levitin, D. J., Chafe, C. H., Berger, J., and Menon, V. (2007). Neural dynamics of event segmentation in music: Converging evidence for dissociable ventral and dorsal networks. *Neuron, 55*: 521–532.

Starr, L. (2011). *George Gershwin*. New Haven: Yale University Press.

Stegemoller, E. L., Skoe, E., Nicol, T., Warrier, C. M., and Kraus, N. (2008). Music training and vocal production of speech and song. *Music Perception, 25*(5): 419–428.

Steinbeis, N. and Koelsch, S. (2007). Shared neural resources between music and language indicate semantic processing of musical tension-resolution patterns. *Cerebral Cortex, 18*: 1169–1178.

Steinbeis, N., Koelsch, S., and Sloboda, J. A. (2006). The role of harmonic expectancy violations in musical emotions: Evidence from subjective, physiological, and neural responses. *Journal of Cognitive Neuroscience, 18*(8): 1380–1393.

Stephan, H., Frahm, H. D., and Baron, G. (1981). New and revised data on volumes of brain structure in insectivores and primates. *Folia Primatologica, 35*: 1–29.

Stephan, K. M., Fink, G. R., and Passingham, R. E. (1995). Functional anatomy of the mental representation of upper extremity movements in healthy subjects. *Journal of Neurophysiology, 73*: 921–924.

Sterelny, K. (2007). Social intelligence, human intelligence and niche construction. *Philosophical Transactions of the Royal Society of London B, 362*: 719–730.

Sterling, P. (2004). Principles of allostasis. In J. Schulkin (ed.), *Allostasis, Homeostasis and the Costs of Physiological Adaptation* (pp. 17–64). Cambridge: Cambridge University Press.

Sterling, P. C. and Eyer, J. (1988). Allostasis: A New Paradigm to Explain Arousal Pathology. In S. Fisher and J. Reason (eds.), *Handbook of Life Stress Cognition and Health*. U.S. and Canada: John Wiley and Sons, Ltd.

Stewart, L. (2005). A neurocognitive approach to music reading. *Annals of the New York Academy of Sciences, 1060*: 377–386.

Stewart, L. (2006). Music and the brain: Disorders of musical listening. *Brain, 129*: 2533–2553.

Stewart, L. (2008). Fractionating the musical mind: Insights from congenital amusia. *Current Opinion in Neurobiology, 18*: 127–130.

Stewart, L., Henson, R., Kampe, K., Walsh, V., Turner, R., and Frith, U. (2003). An fMRI study of musical literacy acquisition. *Annals of the New York Academy of Sciences, 999*: 204–208.

Strand, F. L. (1999). *Neuropeptides: Regulators of Physiological Processes*. Cambridge, MA: MIT Press.

Stringer, C. B. (1992). Reconstructing recent human evolution. *Philosophical Transactions of the Royal Society of London B, 337*: 217–224.

Stringer, C. B. and Andrews, P. (1988). Genetic and fossil evidence for the origin of modern humans. *Science, 239*: 1263–1268.

Sullivan, J.W.N. (1927/1955). *Beethoven: His Spiritual Development*. New York: Vintage.

Sunberg, J. and Lindblom, B. (1976) Generative theories in language and music descriptions. *Cognition, 4*: 99–122.

Suskin, S. (2010). *Show Tunes*. Oxford: Oxford University Press.

Swanson, L. W. (2000). Cerebral hemisphere regulation of motivated behavior. *Brain Research, 886*: 113–164.

Swanson, L. W. (2003). *Brain Architecture*. Oxford: Oxford University Press.

Swanson, L. W. and Petrovich, G. D. (1998). What is the amygdala? *Trends in Neuroscience, 21*: 323–331.

Sylvan, R. (2002). *Traces of the Spirit*. New York: New York University Press.

Tallis, R. (2010). *Michelangelo's Finger: An Exploration of Everyday Life*. New Haven: Yale University Press.

Tarasti, E. (1994). *A Theory of Musical Semiotics*. Bloomington: Indiana University Press.

Tarassti, E. (2000). *Existential Semiotics*. Bloomington: Indiana University Press.

Taruskin, R. (1991). *Stravinsky and the Russian Tradition*. Berkeley: University of California Press.

Tattersall, I. (1993). *The Human Odyssey: Four Million Years of Human Evolution*. New York: Prentice Hall.

Tattersall, I. and Schwartz, J. (2001). *Extinct Humans*. Boulder: Westview Press.

Temperley, D. (2001). *The Cognition of Basic Musical Structures*. Cambridge, MA: MIT Press.

Temperley, D. (2004). Bayesian models of musical structure and cognition. *Musicae Scientiae, 8*: 175–205.

Temperley, D. (2007). *Music and Probability*. Cambridge, MA: MIT Press.

Temperley, D. (2009). A unified probabilistic model for polyphonic music analysis. *Journal of New Music Research, 38*(1): 3–18.

Thaut, M. H. (2005). The future of music in therapy and medicine. *Annals of the New York Academy of Sciences, 1060*: 303–308.

Thaut, M. H. (2010). How Music Helps to Heal the Injured Brain: Therapeutic Use Crescendos Thanks to Advances in Brain Science, by Michael H. Thaut, PhD, and Gerald C. McIntosh, MD. *Cerebrum*. Online from http://ww.dana.org/assets/0/16/32/492/bd99578eae69440c89b1bd7217550cb5.pdf.

Thaut, M. H., Kenyon, G. P., Schauer, M. L., and McIntosh, G. C. (1999). The connection between rhythmicity and brain function. *IEEE Engineering in Medicine and Biology Magazine, 18*(2): 101–108.

Thaut, M. H., Mcintosh, K. W., McIntosh, G. C., and Hoember, V. (2001). Auditory rhythmicity enhances movement and speech motor control in patients with Parkinson's disease. *Functional Neurology, 16*(2): 163–172.

Thayer, A. W. (1871/1920/1973). *Beethoven*. Princeton: Princeton University Press.

Thomas, E. (2001). Empathy and consciousness. *Journal of Consciousness Studies, 8*: 1–35.

Thompson, W. F., Cuddy, L. L., and Plaus, C. (1997). Expectancies generated by melodic intervals: Evaluation of principles of melodic implication in a melody-completion task. *Perception and Psychophysics, 59*: 1069–1076.

Thompson, R. R., George, K., Walton, J. C., Orr, S. P., and Benson, J. (2006). Sex-specific influences of vasopressin on human social communication. *Annals of the New York Academy of Sciences, 103*: 7889–7894.

Thorpe, S.K.S., Holder, R. L., and Crompton, R. H. (2007). Origin of human bipedalism as an adaptation for locomotion on flexible branches. *Science, 316*: 1328–1331.

Thorpe, W. H. (1974). *Animal Nature and Human Nature*. New York: Doubleday.

Thorton-Wells, T. A., Cannistraci, C. J., Anderson, A. W., Kim, C. Y., Eapen, M., Gore, J. C., et al. (2010). Auditory attraction: Activation of visual cortex by music and sound in Williams syndrome. *American Journal on Intellectual and Developmental Disabilities, 115*(2): 172–189.

Tinbergen, N. (1951). *The Study of Instinct*. New York: Oxford University Press.

Tobler, P. N., Fiorillo, C. D., and Schultz, W. (2005). Adaptive coding of reward value by dopamine neurons. *Science, 307*: 1642–1645.

Tomasello, M. (1999). *The Cultural Origins of Human Cognition*. Cambridge, MA: Harvard University Press.

Tomasello, M. and Carpenter, M. (2007). Shared intentionality. *Developmental Science, 12*: 121–125.

Tomasello, M., Carpenter, M., Call, J., Behne, T., and Moll, H. (2004). Understanding and sharing intentions: The origins of cultural cognition. *Behavioral and Brain Sciences, 28*: 675–735.

Tomasello, M., Kruger, A. C., and Ratner, H. H. (1993). Cultural learning. *Behavioral and Brain Sciences, 16*: 495–552.

Trainor, L. (2008). The neural roots of music. *Nature, 453*: 598–599.

Trainor, L. J. (2005) Are there critical periods for musical development? *Developmental Psychobiology, 46*: 262–278.

Trainor, L. J., and Trehub, S. E. (1994) Key membership and implied harmony in Western tonal music: developmental perspectives. *Perception and Psychophysics, 56*: 125–132.

Trainor, L. J., Tsang, C. D., and Cheung, V.H.W. (2002). Preference for sensory consonance in 2 and 4 month old infants. *Music Perception, 20*: 187–194.

Tramo, M. J., Cariani, P. A., Delgutte, B., and Braida, L. D. (2001). Neurobiological foundations for the theory of harmony in western tonal music. *Annals of the New York Academy of Sciences, 930*: 92–116.

Tramo, M. J., Shah, G. D., and Braida, L. D. (2002). Functional role of auditory cortex in frequency processing and pitch perception. *Journal of Neurophysiology, 87*(1): 122–139.

Trehub, S. E. (2001). Musical predispositions in infancy. *Annals of the New York Academy of Sciences, 930*: 1–16.

Trehub, S. E. (1990). Human infants' perception of auditory patterns. *International Journal of Comparative Psychology, 4*: 91–97.

Treitler, L. (1989). *Music and the Historical Imagination*. Cambridge, MA: Harvard University Press.

Tremblay, L., Hollerman, R., and Schultz, W. (1998). Modifications of reward expectation-related neuronal activity during learning in primate striatum. *Journal of Neurophysiology, 80*: 964–977. With permission from the American Physiological Society.

Trimble, M. R. (2007). *The Soul in the Brain*. Baltimore: Johns Hopkins University Press.

Tulving, E. and Craik, F.I.M. (2000). *The Oxford Handbook of Memory*. Oxford: Oxford University Press.

Ullman, M. T. (2001). A neurocognitive perspective on language: the declarative procedural model. *Nature Neuroscience, 9*: 266–286.

Ullman, M. T. (2004). Is Broca's area part of a basal ganglia thalamocortical circuit? *Cognition, 92*: 231–270.

Ullman, M. T., Corkin, S., Coppola, M., Hickok, G., Growdon, J. H., Koroshetz, W. J., et al. (1997). A neural dissociation within language: Evidence that the mental dictionary is a part of declarative memory, and that grammatical rules are processed by the procedural system. *Journal of Cognitive Neuroscience, 9*: 266–286.

Ullman, M. T., and Pierpont, E. I. (2005). Specific language impairment is not specific to language: The procedural deficit hypothesis. *Cortex, 41*: 399–433.

Ungerleider, L. G. and Mishkin, M. (1982). Two cortical visual systems. In D. Ingle, M. Goodale, and R. Mansfield (eds.), *Analysis of Visual Behavior* (pp. 549–586). Cambridge, MA: MIT Press.

Uvnäs-Moberg, K. (1998). Oxytocin may mediate the benefits of positive social interaction and emotions. *Psychoneuroendocrinology, 23*: 819–835.

Unyk, A. M., Trehub, S. E., Trainor, L. J., and Schellenberg, E. G. (1962). Lullabies and simplicity: A cross-cultural perspective. *Psychology of Music, 20*: 308–319.

Van Essen, D. C. (2005). Corticocortical and thalamocortical information flow in the primate visual system. *Progress in Brain Research, 149*: 173–185.

Van Essen, D. C., Anderson, C. H., and Felleman, D. J. (1992). Information processing in the primate visual system: An integrated systems perspective. *Science, 255*: 419–422.

Van Praag, H., Christie, B. R., Sejnowski, T. J., and Gage, F. H. (1999). Running enhances neurogenesis, learning, and long-term potentiation in mice. *Annals of the National Academy of Sciences, 96*: 13427–13431.

Varela, F. J., Thompson, E., and Rosch, E. (1991). *The Embodied Mind*. Cognitive Science and Human Experience Series. Cambridge, MA: MIT Press.

Veenema, A. and Neumann, I. D. (2008). Central vasopressin and oxytocin release: regulation of complex social behaviors. *Progress in Brain Research, 70*: 261–27.

Von Bekesy, G. (1970). Improved musical dynamics by variation of apparent size of sound source. *Journal of Music Theory, 14*: 141–164.

Von Frisch, K. (1953) *The Dancing Bees*. New York: Harcourt, Brace Jovanovich.

Voorhuis, T.A.M., Kiss, J. Z., De Kloet, E. R., and de Wied, D. (1988). Testosterone-sensitive vasotocin-immunoreactive cells and fibers in the canary brain. *Brain Research, 442*: 139–146.

Voorhuis, T.A.M., De Kloet, E. R., and De Wied, D. (1991). Effect of a vasotocin analog on singing behavior in the canary. *Hormones and Behavior, 25*: 549–559.

Vuust, P., Ostergaard, L., Pallesen, K. J., Bailey, C., and Roepstorff, A. (2009). Predictive coding of music-brain responses to rhythmic incongruity. *Cortex, 45*: 80–92.

Vuust, P. and Kringelback, M. L. (2010). The pleasure of music. In M. L. Kringelbach and K. C. Berridge (eds.), *Pleasures of the Brain* (pp. 255–269). Oxford: Oxford University Press.

Vygotsky, L. S. (1934/1979). *Language and Thought*. Cambridge, MA: MIT Press.

Wagner, R. (2011). *My Life*. New York: Tudor Publishing.

Wallin, N. L., Merker, B., and Brown, S. (2000). *The Origins of Music*. Cambridge, MA: MIT Press.

Walker, F. (1962/1982). *The Man Verdi*. Chicago: University of Chicago Press.

Wan, C. Y., Ruber, T., Hohmann, A., and Schlaug, G. (2010). The therapeutic effects of singing in neurological disorders. *Music Perception, 27*: 287–295.

Wan, C. Y., Demaine, K., Zipse, L., Norton, A., and Schlaug, G. (2010). From music making to speaking: engaging the mirror neuron system in autism. *Brain Research Bulletin, 82*: 161–168.

Wang, Z. (1995). Species differences in the vasopressin-immunoreactive pathways in the bed nucleus of the stria terminalis and medial amygdaloid nucleus in prairie voles (*Microtus ochrogaster*) and meadow voles (*Microtus pennsylvanicus*). *Behavioral Neuroscience, 109*: 305–311.

Wang, Z. and De Vries, G. J. (1993). Testosterone effects on paternal behavior and vasopressin immunoreactive projections in prairie voles (*Microtus ochrogaster*). *Brain Research, 631*: 156–160.

Wang, Z., Moody, K., Newman, J. D., and Insel, T. R. (1997). Vasopressin and oxytocin immunoreactive neurons and fibers in the forebrain of male and female common marmosets (*Callithrix jacchus*). *Synapse, 27*: 14–25.

Weaver, I. C., Cervoni, N., Champagne, F. A., D'Alessio, A. C., Sharma, S., Seckl, J. R., et al. (2004). Epigenetic programming by maternal behavior. *Nature Neuroscience, 7*: 847–854.

Weber, C. (1931). An experiment at Wells. *Journal of Higher Education, 2*: 298–304.

Weeks, R., Horowitz, B., Aziz-Sultan, A., Tian, B., Wessinger, C. M., Cohen, L. G., et al. (2000). A positron emission tomographic study of auditory localization in the congenitally blind. *Journal of Neuroscience, 20*: 2664–2672.

Wehr, T. A., Moul, D. E., Barbato, G., Giensen, H. A., Seidel, J. A., Barker, C., et al. (1993). Conservation of photoperiod-responsive mechanisms in humans. *American Journal of Physiology, 265*: 846–857.

Weinberger, N. M. (2004). Music and the brain. *Scientific American, 291*: 88–95.

Weissman, D. (2000). *A Social Ontology*. New Haven: Yale University Press.

Wengenroth, M., Blatow, M., Bendszus, M., and Schneider, P. (2010). Leftward lateralization of auditory cortex underlies holistic sound perception in Williams syndrome. *PLoS ONE, 5*: e12326.

Wernicke, C. (1874). *Der aphasische symptomencomplex: Eine psychologische studie auf anatomischer basis*. Breslau: Kohn und Weigert.

Wheeler, M. (1995). *Reconstructing the Cognitive World*. Cambridge, MA: MIT Press.

Whitehead, A. N. (1919/1982). *An Enquiry Concerning the Principles of Natural Knowledge*. New York: Dover Press.

Whitehead, A. N. (1929/1958). *The Function of Reason*. Boston: Beacon Press.

Whitehead, A. N. (1938/1967). *Modes of Thought*. New York: Free Press.

Whiten, A. (1997). *Machiavellian Intelligence II: Evaluations and Extensions*. Cambridge: Cambridge University Press.

Whiten, A. and Van Schaik, C. P. (2007) Evolution of animal cultures and social intelligence. *Philosophical Transactions of the Royal Society of London, 362*: 603–620.

Wilson, M. (2002). Six views of embodied cognition. *Psychonomic Bulletin and Review, 9*: 625–636.

Wingfield, J. C., Jacobs, J. D., Soma, J., Maney, D. L., Hunt, K., Wisti-Peterson, D., et al. (1999). Testosterone, aggression and communication: Ecological bases of endocrine phenomenon. In M. D. Hauser and M. Konishi (eds.), *The Design of Animal Communication* (pp. 255–283). Cambridge: Cambridge University Press.

Winkler, I., Haden, G. P., Ladinig, O., Sziller, I., and Honing, H. (2009). Newborn infants detect the beat in music. *Proceedings of the National Academy of Sciences of the United States, 106*: 2467–2471.

Winslow, J. T., Carter, C. S., Harbaugh, C. R., and Insel, T. R. (1993). A role for central vasopressin in pair bonding in monogamous prairie voles. *Nature, 365*: 545–548.

Wiora, W. (1965). *The Four Ages of Music*. New York: Norton.

Wise, S. P. (1985). The primate premotor cortex: Past, present and preparatory. *Annual Review of Neuroscience, 8*: 1–19.

Wise, S. P. (2006). The ventral premotor cortex, corticospinal region C, and the origin of primates. *Cortex, 42*: 521–524.

Wittgenstein, L. (1953/1968). *Philosophical Investigations*. New York: MacMillan.

Wong, P.C.M. and Diehl, R. L. (1999). The effect of reduced tonal space in Parkinsonian speech on the perception of Cantonese tones. *Journal of the Acoustical Society of America, 105*: 1246.

Wong, P.C.M., Skoe, E., Russo, N. M., Dees, T., and Kraus, N. (2007). Musical experience shapes human brainstem encoding of linguistic pitch patterns. *Nature Neuroscience, 10*: 420–422.

Wood, B. A. (1992). Origin and evolution of the genus homo. *Nature, 355*: 783–790.

Wood, B. A. (2000). The history of the genus homo. *Human Evolution, 15*: 39–49.

Wynn, T. and Coolidge, F. L. (2008). A stone-age meeting of minds. *American Scientist, 96*: 44–51.

Young, A. W. (1998). *Face and Mind*. Oxford: Oxford University Press.

Young, L. J., Nilsen, R., Waymire, K. G., MacGregor, G. R., and Insel, T. R. (1999). Increased affiliative response to vasopressin in mice expressing the V1a receptor from a monogamous vole. *Nature, 400*: 766–768.

Young, L. J., Toloczko, D., and Insel, T. R. (1999). Localization of vasopressin (V1a) receptor binding and mRNA in the rhesus monkey brain. *Journal of Neuroendocrinology, 11*: 291–297.

Zak, P. L. (2008). The neurobiology of trust. *Scientific American, 298*: 88–95.

Zatorre, R. J. (2001). Neural specializations for tonal processing. *Annals of the New York Academy of Sciences, 930*: 193–210.

Zatorre, R. J., Belin, P., and Penhune, V. B. (2002). Structure and function of auditory cortex: music and speech. *Trends in Cognitive Sciences, 6*: 37–46.

Zatorre, R. J., Chen, J. L., and Penhune, V. B. (2007). When the brain plays music: Auditory-motor interactions in music perception and production. *Nature Publishing Group, 8*: 547–558.

Zatorre, R. J. and Halpern, A. R. (2005). Mental concerts: Musical imagery and auditory cortex. *Neuron, 47*: 9–12.

Zatorre, R. J. and Krumhansl, C. L. (2002). Mental models and musical minds. *Science, 298*: 2138–2139.

Zbikowski, L. M. (1998). Metaphor and music. *Society for Music Theory, 4*(1): 502–524.

Zbikowski, L. M. (2005). *Conceptualizing Music*. Oxford: Oxford University Press.

Zbikowski, L. M. (2011a). Musical gesture and musical grammar: A cognitive approach. In A. Gritten and E. King (eds.), *New Perspectives on Music and Gesture* (pp. 83–98). Burlington, VT: Ashgate Publishing.

Zbikowski, L. M. (2011b). Music and movement: A view from cognitive science. *Bewegungen Zwischen horen und Sehen.*

Zeigler, H P. (ed.) (2008). *Neuroscience of Birdsong.* Cambridge: Cambridge University Press.

Zeki, S., Watson, J. D., and Frackowiak, R. S. (1993). Going beyond the information given: The relation of illusory visual motion to brain activity. *Proceedings of the Royal Society of London–Series B, Biological Sciences, 252*: 215–222.

Zentner, M. (2011). Homeric prophecy: An essay on music's primary emotions. *Music Analysis, 29*: 102–125.

Zentner, M. and Eerola, T. (2010). Rhythmic engagement with music infancy. *Proceedings of the National Academy of Sciences of the United States, 107*: 5767–5773.

Zentner, M. R. and Kagan, J. (1996). Perception of music by infants. *Nature, 383*: 29–30.

Zimmer, C. (2005). *Smithsonian Intimate Guide to Human Origins.* Washington, DC: Smithsonian Books.

Zink, C. F., Kempf, L., Hakimi, S., Rainey, C. A., Stein, J. L., and Meyer-Lindenberg, A. (2011). Vasopressin modulates social recognition-related activity in the left temporoparietal junction in humans. *Translational Psychiatry, 1*: e3.

Zitzer-Comfort, C., Reilly, J., Korenberg, J. R., and Bellugi, U. (2010). We are social—therefore we are. In C. Worthman, et al. (eds.), *The Interaction of Care Given, Culture, and Developmental Psychobiology* (pp. 136–166). Cambridge: Cambridge University Press.

Zoidis, A. M., Smultea, M. A., Frankel, A. S., Hopkins, J. L., Day, A., McFarland, et al. (2008). Vocalizations produced by humpback whale (*Megaptera novaeangliae*) calves recorded in Hawaii. *Journal of the Acoustical Society of America, 123*: 1737–1746.

Index